Nothing and Nothingness:

Toward an Apophatic Science

Gustavo Candia and Rodney J. Parrott

ISBN: 978-1-777024208

Front cover image by Gustavo Candia

Book design by Bookwrangler

Printed by Nihiliverse, in Toronto, Canada

FIRST EDITION

www.nihilology.com

Contents

Introduction

In this phase of life, nothing is more essential to us than academic career, professional respect and reasoned science.

We have labored with joy over the past decade to give birth to a meaningful critique of, and guidance for, popularizers of science from the vantage point of Nothing and nothingness. This book is the child of our labor.

We begin by introducing our readers to the popularizers, or POPs as we affectionately call them. They hold Ph.D. degrees in physics and cosmology. They act as the scions of accessible wisdom regarding the nature of the universe and they function as POPstars of the intellect, establishing the mythos of our contemporary science-based culture. They are publishing books in huge volumes and are becoming widely viewed YouTube stars, TV personalities, and newsworthy cultural personae.

In order to speak to their literate fans, all popularizers of science forego their esoteric language of mathematics and thereby, forced to resort to words and images, have become story tellers. As popularizer scientists have become embedded in our culture as the ultimate explainers of the creation and workings of the universe, their stories have also evolved into the scriptures of western culture.

In becoming personal authors of their cosmic stories, POPs have re-introduced the human being back into their canon. Thus, by definition, they have chiseled a chink in scientific objectivity, in the bedrock of the scientific worldview, allowing the emotional, believing, subjective and narrative components of the human being, once again, to be an integral aspect of the cosmos.

Further, our critique includes the perspective that popularizer scientists are, in an array of ways, usurping the traditional roles of religion in our new scientific culture of the West. For instance, ever since Einstein introduced the world to relativity, rocketing the picture of the universe well beyond the ken of literate laypeople, science has served as a comforting belief system whose stories provide us with a vision of the manifest universe to believe in. Indeed, science became its own "ism." Yet POPs are blind to the fact of their own self-created scientism.

Also, with the search for a single essence to the universe and the search for a unified theory of the universe, popularizer scientists exhibit quests utterly parallel to those of religious monism on the one hand and religious monotheism on the other. And finally, POPs have faith in the belief that the universe is a single unity and that beauty will serve to guide them to discover that unity. POPs indeed have faith and, ironically enough, it aids them in being better scientists!

In the end, we suggest that popularizers of science don't need us to argue them into accepting religious approaches to understanding the world we live in. They are already doing religion!

They are already doing philosophy as well, all their shouted denials aside. Traditionally, POPs have focused on the "how" of the universe, the **mechanics** of it all. Yet, with the ever-growing introduction—and validation—among the POPs of the "why?" question, science is widening its scope to incorporate the quest for **meaning** in our manifest universe. POPs science now, by its very nature, is its own form of philosophizing. We proclaim that the POPs are in fact today's torch bearers of Existentialism and the misguided incarnations of Nihilism's übermensch.

It would not be fair for us as authors to simply critique the POPs without offering a way to evolve their science. And indeed, we do suggest a rational strategy for popularizers of science called Nihilology and ask that they accept the input of Nihilology toward the improvement of their theorizing and empirical research. This evolved form we call "apophatic science."

The apophatics held that lying just beyond all the continuums of the world—beyond materiality, beyond personality, beyond rationality—lies Nothing which is not a thing, is not a person, is not describable.

We would like to suggest the popularizers of science should declare Nothing and then heed the research paradigm as put forth by Nihilology. We ask POPs to take on an epistemic approach that differentiates between a problem solvable in the world of rational reach and a mystery accepted as beyond reason's reach. That they embrace the challenge of solving problems about the universe and embrace the unsolvable Nothing of the universe. This embrace of the rational and the extra-rational is indeed the rational thing to do. This embrace will bring about a reflective consciousness regarding POPs übermensch personae, mollify their hubristic belief in their rational capacity and improve their science as well.

Now, what do we intend with the title of our book: *Nothing and nothingness*? The history of science over the past four hundred years has witnessed a struggle, taken on by the saints of the popular science canon from Galileo Galilei to Michael Faraday to Albert Einstein. Its goal was to grasp a rational, empirical handle on Nothing and has led to stunning contributions in the fields of mathematics, physics, and psychology. Indeed, Nothing has evolved into a guiding notion for science researchers.

Nonetheless, after four hundred years of searching, most popularizers of science and historians of science agree that seeking Nothing within the realm of the material world is in vain. As we say, "Nothing is not." These scientists have done a thorough job of pursuing the Nothing of materiality and found it not. "Nothing is not a something." The objective world has been plummeted, and each and every time, in each and every realm of the manifest scanned, Nothing always turns out to be elusive. The sought-after, empty state is actually nothingness, not Nothing.

This conclusion is in accord with the teachings of Nihilology: on a continuum ranging from heavy materiality of stars and planets to utterly light materiality of unstable space and virtual particles, Nothing is not to be found. Nothing lies off the ends of the continuum of materiality all together. Going one step farther, Nihilology also holds that Nothing lies outside the realm of rationality. Nothing lies over the cliff edge of all continua of rational knowledge and of experience itself. Our call to scientists is, at the very least, to **believe** in Nothing.

Now, let us take a moment to tell our readers one aspect of how the book came to be. We hold there is one foundational insight—"outsight" as we say—that every philosopher, religionist or scientist must undergo in order to bring a new and significant interpretive tool to our library of human knowledge about the reality of the manifest world: the outsight of all-pervasiveness.

For instance, the nineteenth century philosopher Friedrich Nietzsche saw a long line of chinks in the worldview of nineteenth century Europe. This series of seeings led him to the outsight that the entire worldview of his era was throughout moribund, "God is dead." Or the sixth century BCE religious leader known as Buddha, who saw that human suffering has its root in many faulty experiences of selfhood. Then he suddenly awakened to the pervasive fact that indeed there exists no such thing as selfhood at all. Or, Frank Wilczek, a contemporary MIT theoretical physicist, began to see evidence of a "miraculous harmony" existing between the mind and matter and that this harmony is founded on beauty. Wilczek then came to the outsight that beauty is not simply a guide to discovery of the physical laws of the universe; rather beauty indeed constitutes these laws: beauty is everywhere and is in fact inherent to the nature of everything!

These outsights provided a heuristic capacity—a tool for interpretation—enabling these innovators in philosophy, religion, and science to transform our understanding of the world. However, this outsight of pervasiveness need not be essential only for the philosophical, religious, or scientific elites. It is necessary in the community and personal lives of our readers as well. Say you experience a series of biased slights due to your skin color. Then an a-ha! occurs to you one day that indeed bias permeates your culture and is one of its dynamic foundations. This a-ha! guides you through activism for social justice.

Or you may have a child that you adore and treat with the greatest care and nurturing. Then it dawns upon you that all positive human relations are founded on love and that love is the very definition of relationship. This dawning serves as a beacon in your striving for more heartfelt engagement among all people.

A foundational outsight of all-pervasiveness also occurred for us. It gave birth to this book and to a new school of philosophy of science called Nihilology. What began as a critique of David Hume's views on creativity from Nothing grew into a critique of Lawrence Krauss's views on creation from Nothing. Then to the realization that writers of popular science generally don't know Nothing and if they did their science would be better. Finally, the outsight crystalized. Nothing is everywhere and its proper understanding will revolutionize our notions of the existent cosmos. This is our subtlety, our potent heuristic, our contribution to the library of human knowledge. We are good at sensing the all-pervasiveness of Nothing and good at sensing when writers of popular science do not.

Perhaps we should also take a moment to introduce two basic axioms that guide the methodology of Nihilology. First, it is in the pattern of knowledge, gleaned from the teachings of religion, science, and philosophy, that we find the most powerful tool for solving the conundrums of the manifest universe. For example, from the teachings of religion, East and West, we learn that Nothing and creation are eternally present. Creation is not the end of Nothing. Nothing is not lost at the creation of the universe. Nothing gives creation what we term an "isometric balance" that allows created things to function.

Second, we, as writers, adhere utterly to a regimen of technical language that forgoes any facile joke-making regarding the multi-valent connotations of the term "Nothing". Throughout our book "Nothing" is a declarative, uttered with the intent of coaxing our readers toward a minimalist understanding of Nothing's nature as one of verity and gravity. A Nothing that resides in our readers as true with a pull to it, free of any inherent characteristics or ideational handles.

With all this in mind, as we enter into the seven sections of our book, let us say clearly: There is no pith to Nothing! We can't reveal the essence of Nothing in this introduction. We are so sorry. As readers you will not be able to peruse this introduction, get the pith, then head out to the classroom to lecture about Nihilology. You will not be able to go to a party and hold forth on what you learned about Nothing from our book.

Rather, we invite you to read at your own pace, but read the whole book. In the course of doing so, perhaps over a period of days or weeks, in reading the final page and putting the book down, you will have come to a deep appreciation of Nothing and how powerful a heuristic it is. Your thoughts on the world will have evolved.

(1)
Why Nihilology?

Nothing Bursts Upon Us

In reflecting upon our intellectual history, we find the birth of this book, and our life project, in a paper Gustavo wrote in 2005 while a graduate student at the Institute of History and Philosophy of Science and Technology at the University of Toronto. The title of the paper was "Causation from Nothing." It took the great eighteenth century Scottish philosopher David Hume to task regarding his thoughts on the nature of causation. So many respected thinkers of Hume's time and since have given their critique on this subject; it did show a bit of hubris on Gustavo's part as a young and green history of philosophy student to join this august fray. (The professor, perhaps, showed his agreement in this sentiment by granting Gustavo a B+.) Here is the heart of what he argued way back then:

> The focus of this paper is to explore the ramifications of one of Hume's less familiar points about causation: the possibility of something being caused by nothing. Hume

> suggests that causal laws are not necessary because we can imagine something having *no cause*. The main difficulty Hume confronts with this reasoning is: even if one could *imagine* something having no cause, this would not imply that one can *logically* assert such a point. Hume nor anybody else for that matter, cannot make *logical* claims about there not being a necessary connection between cause and effect by referring to non-causes or events such as "no cause". To solve this problem, I will demonstrate that *nothing must be something*. "Causation by nothing" must be re-interpreted to mean that something is actually caused by nothingness, i.e. the *state* of nothing, which is indeed something.

Forgiving the inchoateness of the ideas before, even until this day, we hold that Nothing cannot cause something to come to be; only nothingness has the capacity to do so. But we will have this discussion later in the book when we state clear definitions and bring forth our comprehensive perspective. For now, it is enough to note this paper as the birthplace of our decade-long investigation of Nothing.

This investigation aligns us with thinkers from at least the past 2,500 years. Around 400 BCE, the Greek philosopher Democritus, following in the sandal steps of his mentor Leucippus, posited that the universe is constituted of eternal uncuttable atoms. The atoms moved about in an equally eternal, infinite void. Democritus was the first person we know of to posit a real existence of Nothing. Today's astrophysicists often call Democritus the father of modern science in that he saw the universe as a material phenomenon, with no need of the gods. (Some also admire him as the "Laughing Philosopher" because of his emphasis upon cheerfulness in a world devoid of the religious.)

Around 300 BCE, the world's first zero appeared in Babylonia. It wasn't a round circle then, rather a pair of left-slanted arrows that began to show up in a base 60 system for tracking the heaven's constellations. Around 100 BCE, Buddhists of the Mahayana school advanced the notion that the entire manifest universe is actually an emptiness. Neither the human being nor the objects of the world have any independent existence.

It is an honor for us to rub minds with some of the great thinkers of history in our zealous search for understanding Nothing. For instance, we feel a kinship with K.C. Cole, the remarkable science writer for the *Los Angeles Times* and professor at USC Annenberg School for Communication and Journalism. In her book *The Hole in the Universe* she exclaims: "...nothing may be the single most prolific idea ever to plop into the human brain."

Beginning with Galileo Galilei in the 1600s, who proved with his amazing telescopes that the earth was *not* the hub of the universe and who held to this emptying of the center in spite of the Inquisition declaring him "vehemently suspect of heresy." To Michael Faraday in the 1800s, who built on the then-new discovery that atoms of matter are 99.99% empty space by declaring that stuff is *not* stuff at all, but rather consists of fields of force in space. To Albert Einstein in the 1900s, who proved dramatically that space is *not* the empty medium of force fields, but rather is a space-time continuum that constitutes the very nature of force fields themselves. The study of what is *not* has been a marching forth of knowledge that over and over again pulled the rug out from under the feet of us as individual people, as intuitive humans, and as cultures. What we taken for granted repeatedly turned out *not* to be.

The struggle that has taken place over the past four hundred years to get a mental handle on Nothing has drastically changed the fields of mathematics, physics and psychology. Scientists looked deep into the *not* and saw a rich treasure. Again, according to K.C. Cole, "the evolution of nothing into a full-fledged player in the universe stands as one of the single greatest paradigm shifts in human thought."

Ten years ago, the rug was indeed pulled out from under our intellectual feet. We had been guided by the accepted academic notion that history of science was to be understood by looking at the evolution of hypotheses, experimentation, and theory. Each new theory building positively upon the last. Then we discovered that this academic guide, while simple and comforting, was in fact misguided. Science, at least the important kind, has been a history of dismantling, not building up. This was a tectonic paradigm shift for us. And, we shall argue, Nihilology will assume its place in history as yet another dismantling… contributing to the evolution of scientific understanding of our world.

Let us quote Cole one more time, "While no one was looking, nothing became a central player that creates number systems out of whole cloth; bubbles up matter and universes; materializes sights, sounds, perceptions." *While no one was looking* is the key phrase here. What happened ten years ago is that we began to look closely at the history of science, and we suddenly began to see Nothing everywhere.

Jeremy Webb, editor of the New Scientist Book, *Nothing: From Absolute Zero to Cosmic Oblivion*, goes straight to the pith of the intellectual power of our insight, "In this way, nothing becomes a lens through which we can explore the universe around us and even what it is to be human." Robert Kaplan also agrees with us. In his book *The Nothing that is: A Natural History of Zero,* he exclaims, "If you look at zero you see nothing; but look through it and you will see the world." Zero, though empty, reveals much!

This seeing of Nothing everywhere has steeped in us for the past ten years. The resulting brew is Nihilology: a fully elaborated, subtle, and sophisticated study of Nothing. Nihilology is a potent tool, a heuristic that aids us in understanding many fields, e.g., cosmology, psychology, religion. We can employ Nihilology to guide us to deep interpretations. Once we see Nothing as the ultimate heuristic, we can use it as a rule of thumb, as a sharp cut—not a *short*cut—to reveal much.

Rodney once wrote a novel about the life of a wanderer. It told of the wanderer's economics that placed no causal link between work performed and wages received; of the wanderer's romance that was, of necessity, momentary; of the wanderer's purpose in life to perform ethical illusions for the benefit of those whose lives he encountered en route. These individual escapades were very engaging to write unto themselves, but as he began to compose the last chapter of the novel, he realized all of a sudden that the overall book was in fact a comprehensive analysis of life from the viewpoint of movement! And this revealed much about his own life from that non-vantage point. Nihilology is much the same. Once we understand it as a complete analysis of the universe from the viewpoint of Nothing, it reveals much.

And who, as readers of our book on Nihilology, will benefit the most? Our hoped-for audience, the readers we have in our mind's eye as we write, are scientists, especially popularizers of physics. We affectionately tag them the POPS There is an especially urgent need for Nihilology among those astrophysicists and cosmologists who, without discipline, toss about notions of Nothing as they popularize leading edge ideas about how the universe came to be and the essential nature of the manifested world. We also hope to draw the attention and respect of the ever-growing readership of POPs writing. Indeed, it is in the employment of Nihilology that these intelligent lay readers will gain a deeper understanding and appreciation of the popular science they so avidly study.

Be Afraid. Be Very Afraid

Nothing has taken on many guises over the past 2,500 years. It has been called infinity, chaos, the void, aether, and dark energy. One thing that brings these guises into one unified mask is fear. Thinkers have cowered while looking at the mask of Nothing. Whole cultures have made it unlawful to even look at the mask. At the end of the thirteenth century, for example, the city of Florence, Italy, banned the use of zero, recently imported from the Arab world. They feared too much fraud would occur among merchants if they could increase a number's value by quietly adding a digit on the end. (They could not pull this ruse with the familiar, non-positional system of Roman numerals.)

In fact, the very idea of Nothing inspired absolute terror in thinkers of Classical Greece, epitomized by Aristotle. As so poignantly stated by Brian Rotman in his book *Signifying Nothing,* "The prospect of an unclassifiable emptiness, an attributeless hole in the natural fabric of being, presented itself as a dangerous sickness, a God-denying madness that left him [Aristotle] with an ineradicable *horror vacuii.*"

Around 400 BCE, the Greeks took on a new and agile numbering system that used letters: *alpha* + *alpha* = *beta* (1 + 1= 2). They did not yet have zero, so they didn't have to contend with the *nil* repulsion like the later Europeans did. However, they could still calculate complex sums, employing other letters for ten and beyond, *omicron* for 70 and *psi* for 700. With this new system, they dealt very well with large series of numbers, but infinity itself was quite a different matter. Aristotle, in the third book of his famous *Physics,* holds that infinity cannot exist as an actuality in any way, not as a body, not as a substance, and not as a void. The term Aristotle used for infinity was *apeiron.* This was a word without any of the positive connotations and feelings we have for the infinite today. (Buzz Lightyear inspired us to go to "infinity and beyond!") As Brian Clegg states in his book *A Brief History of Infinity: The Quest to Think the Unthinkable,* "It meant without bounds, out of control and untidy—*apeiron* was not a pleasant thing. Although it did not have the same denotative meaning, *apeiron* had similar negative associations to those we now give to 'chaos' in normal English…"

We are not overstating the fear that thinkers over long periods of history have experienced when attempting to understand Nothing. The number of scholarly works and popular writings about Nothing with such words as "avoiding," "mysterious," and "terrifying" on their covers attests to this seemingly universal repulsion to what is *not.* We need a vehicle to carry us beyond where our minds are able to go alone, beyond the bounds of mental safety. Nihilology is such a vehicle. It provides structure, an intellectual ground to stand on as we look upon the mask of Nothing.

This fear of Nothing is not just a phenomenon of the ancients; it is alive in later eras as well. Johannes Kepler, perhaps the central figure of the western world in bringing about the scientific revolution, rejected the classic Stoic and Atomist arguments for a universe infinite in extent. And his reasons were, seemingly, psychologically born. He timidly stated, "[The very thought of an] infinity bears some hidden terror, in whatever form; in this immeasurable space we would be all but lost – it has no center and no boundaries and no specific location can be fixed in it."

Kepler believed—and it is indeed a belief—in a finite universe with impenetrable boundaries; his world was protected by a sort of cosmic great wall. Oddly, he used scientific observation to overturn Aristotle's philosophic approach to orbits, which allowed him to discover that planets move in ellipses, not circles, as demanded by Aristotle's axiom of perfect geometry in the heavens beyond the moon. Yet, Kepler continued to cling to Aristotle's *horror vacuii* in his fear of Nothing.

Let us conclude this section with a practical warning: it has proven to be mentally disorienting to contemplate the mask of Nothing. It can be said that two famous scientists have gone mad attempting to contend with infinity. Georg Cantor was the originator of set theory, the great leap forward in the field of mathematics as presented in his famed paper of 1874, "On a Characteristic Property of All Real Algebraic Numbers." His work also influenced the turn-of-the-century debates in the field of philosophy as he presented a proof that infinity is an actuality, not just a potential. Beyond that, he posited that there are different orders of infinity as well; —some infinities are bigger than others! This shocked the imagination of most of his contemporaries. It seems to have shocked Cantor as well, for he had a major nervous breakdown while applying his new-found mathematics to the degrees of infinity. At the end of his life, Cantor could not manage any mathematical contemplations at all with his mental challenges of depression and paranoia. He lived for many years in a sanatorium near the University of Halle in the German province of Saxony where he had been a professor for his entire career.

We might note that Cantor was still working on yet another unsolvable conundrum, this one in the realm of his dear Christian religion. He was trying to prove that Joseph of Arimathea was actually the *natural* father of Jesus, as opposed to the more famous Joseph who, one might say, adopted Jesus. Again, another shock to his contemporaries who believed in the virgin birth.

Another mathematician's mind also succumbed to the contemplation of infinity. Kurt Gödel was plagued by an insatiable drive to understand every aspect of his world. (His Austrian family named him *Herr*

Warum or Mr. Why.) This drive led him to great heights as a professor, as a close friend of Albert Einstein while at Princeton, and to great fame as the genius behind the Incompleteness Theorems.

Gödel carried on Cantor's legacy by bringing forth a proof of Cantor's Continuum Hypothesis. It asserts that every infinite set of real numbers has cardinality. (Cardinality is the number of elements in the set, or its size.) Now, Gödel held a deep confidence that mathematics was the absolute prior science underlying all sciences. This belief, along with his Mr. Why personality, may have led to his demise when he attempted to employ the Continuum Hypothesis, and other set theory principles on the infinity of mathematical objects, to a mathematical proof for the existence of God. He began to develop paranoid symptoms as witnessed in his hiding all theological writings from his colleagues until he knew he was close to death. As he once wrote to his mother, "Ninety percent of contemporary philosophers see their principal task to be that of beating religion out of men's heads." Gödel's greatest paranoia was a debilitating fear of being poisoned. Near the end he ate only food that his wife had cooked. When she was hospitalized, he quit eating altogether and died.

As we see in the cases of Cantor and Gödel, it is dangerous to contemplate the Nothing of infinity, to think bigger than that which we define as the biggest possible thing.

Put Something in There!

Fear has driven us to reject Nothing. As Marcelo Gleiser states in his article "Avoiding The Void: A Brief History Of Nothing(ness)", "Total emptiness, if it exists, has foreboding undertones. Nothingness is scary. We want to fill it up with something." One very effective way to do so is to seek relief by making Nothing into a more manageable mask by filling it up, giving it features of some sort. Aether has been the most long-lived feature. Aether has often been the answer to our search for some prime element in the universe, that which permeates everything or is the basis for everything.

Aether began as the pure air breathed only by the Greek gods in their rarefied cloud palace above Mount Olympus. Aristotle overcame his *horror vacuii* by filling the regions above the moon with a fifth element, a quintessence in addition to the four terrestrial-level elements of air, earth, fire, and water. He called it aether and again gave it properties of purity, this time in the form of pure geometry. Aristotle held that the perfect form was a circle. And, since the planets moved about in the aether, they moved in circles. (This was an admirable attempt at explaining orbits.)

Throughout its history, aether has filled up Nothing for every sort of thinker, from philosophers to theologians, scientists to ethereal spiritualists. (To gain an appreciation of the fringe circles in which aether travels, take a look at this book by William Lyne, *Occult Ether Physics: Tesla's "Ideal Flying Machine" and the Conspiracy to Conceal It* that presents a "factual history" of Nikola Tesla's work on electric-powered flying saucers.) Aether has functioned as a glue-like medium, holding together technology, science, and spiritualism and lending its aroma to all three.

It is very important to note that any time a writer in the area of history of science, history of ideas, or history of philosophy who takes on the task of explaining and elucidating the history of the aether, they face a seeming chaos of mysteries, arcane beliefs, and conspiracies. As Jo Milutis in his book *Ether: The Nothing That Connects Everything* warns us, "The irrational is behind every cloud." This fact of history speaks to the difficulty of our creating Nihilology. It also speaks to the necessity of Nihilology. As we argue below, Nihilology is a complete heuristic of Nothing, a tool for penetrating this notion that stands at the heart of human thought about the nature of the world. Indeed, Nothing gives one access to relative knowledge of everything

With this in mind, let us return to aether, just as it took a central role in the field that is now known as astrophysics. In 1865 James Clerk Maxwell did the field of physics a momentous good deed by bringing together for the first time, electricity, magnetism, and light as manifestations of the same phenomenon: electromagnetic radiation. Maxwell's insight was essentially based on his measurements that electromagnetism was wavelike and it propagates at 300,000 kilometers per second, at the speed of light!

It was already an established scientific fact that mechanical waves, like sound, could only propagate through the air; air serves to "carry" sound energy from one place to another. But electromagnetic waves move through the vacuum of space. Maxwell could confirm this simply by looking up at the heavens at night. He could *see* the light coming to his eyes from the stars.

So, because electromagnetism is a wave, Maxwell naturally inquired "A wave in what?" In his classic 1865 article "A dynamical theory of the electromagnetic Field" he proposed an answer, "If there is no medium for light, how does it move?" A wave in Nothing was impossible in Maxwell's universe, so he filled Nothing. Electricity, magnetism, and light are "transverse undulations of the same medium" and, in a nineteenth century scientific prose that still allowed for the poetic, "affections of the same substance."

Yet, what could this carrier medium possibly be? Newton had proved long before that it could have no density. It had to be transparent, since Maxwell could see the starshine. It had to be of a strong material to allow for the propagation of fast waves, and it had to have zero weight. This medium for undulations and affections came to be known—again in that wonderfully genteel scientific language—as luminiferous aether, or "light-bearing pure air". Thanks to Maxwell, Nothing had been filled again. Alas, aether made its return, this time with the imprimatur of "solid" science.

However, the experimental detection and measurement of aether seemed to be beyond nineteenth century science. Albert Michelson, the Nobel laureate famed for his accurate measurement of the speed of light, turned his professional skills to measuring the luminiferous aether, which he coined as the aether wind. Michelson envisioned the aether "blowing" at a constant speed. And interestingly, for his daughter he also painted a simple picture of his famous Michelson-Morley experiments. Making use of the storyteller's wand to transform the wave quality of aether from gaseous to liquid, he told her that aether was a flowing river. The light he sought to measure was just like swimmers going from shore to shore or up and down stream. Many scientists have responded to this simple childlike story; it has helped them understand the illusive aether as well.

Michelson hypothesized that the speed of light must be relative to the aether wind, just as the speed of sound is relative to its medium, air. Thus, if he could very precisely measure the speed of light traveling upwind and compare it with the speed of light traveling downwind, the difference between the two measurements would be exactly twice the speed of the aether wind. Unfortunately, even with his ingenious machinery created to reflect single color light between mirrors, Michelson found no evidence of aether wind. He tried more and more sophisticated experiments but failed. (Of course, viewed from the evolved science of the twenty-first century, there are plausible ways that the aether could exist while avoiding detection within Michelson's experimental confines. For example, aether drift could be a phenomenon occurring in a fourth spatial dimension, or the aether surrounding an object could be gravitationally attracted to it, thus forming a bubble of static aether blowing about in the wider aether wind.)

Then came Einstein's special theory of relativity in 1905 that effectively explained the motion of light through a vacuum with no need of a medium. It is not that Einstein positively proved that aether did not exist, what he demonstrated is that aether is not a necessary ground for relativity theory. Yet, his discovery doomed aether in the scientific world of the day. Because aether was not measurable and was usurped by relativity as an explanatory tool, Nothing became, once again, empty.

However, thankfully, the story of aether has more chapters. In 1998, astronomers peering through the Hubble Space Telescope at supernovae, concluded that the universe is expanding faster and faster. To explain this phenomenon, they posited that the whole of space is filled up with a mysterious entity which, because nobody understands its essence, they called dark energy. Scientists at the National Aeronautics and Space Administration have suggested it might be a type of "dynamical fluid." This sounds like a contemporary version of Aristotle's pure air, Maxwell's undulating luminiferous medium, and Michelson's cosmic wind. In a sense, dark energy could be thought of as... aether. So, Nothing remained empty for only 90 or so years out of the last 2,500. Today, it is filled up again. We have no need to fear!

Nothing Over the Edge of Continua

Aether is at base a vibrational medium. It is waves and pulsations, whereas infinity is about limits, a non-ending series of sizes. Nothing, as we have explored thus far, is about materiality, a diminution of molecules on down to naught.

Aether is the extent of subtlety, and it extends upward from the moon to the sky to the heavens: liquid, gas, vapor, gaseous medium, and beyond. Infinity is the extent of sequences, the extent of space, the extent of any numbering, and beyond. And so with Nothing. It is the extent of absence. Things, molecules, atoms, particles, nano-particles, quarks, spin. Nothing shrinks down, less and less, and below.

The commonality in all these concepts is that they lie beyond the end of some continuum. This notion of beyond the extent is not new; the great Anselm of Canterbury based his famous definition of God upon this beyond way back in the eleventh century, "Therefore, O Lord, You who give understanding to faith, grant me to understand—to the degree You know to be advantageous—that You exist, as we believe, and that You are what we believe [You to be]. Indeed, we believe You to be something than which nothing greater can be thought." Beyond the extent of air, beyond the extent of numbers, beyond the extent of stuff, beyond the extent of thought.

It has struck us in writing our book that we humans exhibit a yearning to go beyond the extent. And these concepts have allowed us to attempt some sort of thought beyond the extent of what can be conceived. Beyond the extent seems to pull at us to think beyond the daily, and this thinking beyond serves to stretch our mentality itself. As authors, we are convinced, along with Rudy Rucker, who says in his book *Infinity and the Mind—the Science and Philosophy of the Infinite*, "thinking about infinity leads to many fascinating paradoxes. By examining these paradoxes, we learn a great deal about the human mind, its powers and its limitations." Thus, conceiving of the inconceivable expands the bounds of what we know about the world. By contemplating the beyond, we rope a few more bits of understanding into the corral of conceivability and make our mental capacity grow as well.

This is a very important notion. Nothing has changed drastically over the history of theology, philosophy, and science. In each era of history, the specific nature of Nothing has eluded all those who sought it through reflection, yet so much has been learned as a result of pursuing definitive understanding. For example, even though his life-long reflection on infinity broke the bounds of his sanity and added to his deathly mental state, nevertheless, Georg Cantor discovered along the way the potent mathematics of transfinite set theory (Yes, "to infinity and beyond!"), which has greatly benefited mathematical physics. It also has solved many problems in the natural world regarding the chemical properties of light, heat, and electromagnetism. Though Nothing has never been found conceptually, its seeking has produced great knowledge.

In this search, we need a tool to guide us out where common mentality gets lost. Such is the nature of our heuristic: Nihilology. Nihilology is, at base, a tool for contemplating Nothing, a tool for looking out there and making valid conclusions about what we see. This contemplation often begins as rational deduction and evolves into creative imagination or ampliative thinking. Cosmologists and astrophysicists frequently attest that creative imagination, the cerebral cousin of awe, comes about from viewing the universe on a starlit night. As we shall see, this gazing at the stars at night, filled with awe, motivates the scientific search of many an astrophysicist.

To go one step farther, the continued arduous contemplation of Nothing may—for certain scientists or philosophers with probative intent and skill—then bring about a temporary exhaustion of mental concepts. A successful conclusion to a problem may not be an answer at all, rather a relieving cessation of zealous thought on the problem. From the pen of the great Canadian Jesuit priest Bernard Lonergan, who is regarded by many as one of the most important thinkers of the twentieth century, arose the foundational study *Insight: A Study of Human Understanding.* We will speak much of Lonergan later on, but for now listen to this wisdom, "[Insight is] a quite distinct activity of organizing intelligence that places the full set of clues in a unique explanatory perspective. By insight, then, is not meant any act of attention or advertence [cognizing] or memory but the supervening act of understanding." He then follows with, "What we have to grasp is that insight (1) comes as the release to the tension of inquiry, (2) comes suddenly and unexpectedly..."

Contemplating what is beyond the limits serves as a goad to look farther, lower, subtler. The guidance of Nihilology can lead us beyond the limits of thinking. Then, when mental concepts do re-appear, the first thought is filled with energy. As we will discuss soon, Nihilology calls this energetic thought "outsight", the a-ha! moment so sought after by philosophers and scientists alike.

However, for the sake of scientific investigation, "Nothing" must be strictly defined, regardless of its illusive nature and difficult confirmation. Let us begin this task boldly by stating: Nothing does not exist within the realm of human rational comprehension. Yet, it is still true that the contemplation of Nothing serves a positive function for scientists and philosophers. It stretches the mind to find out more.

We have noted several times K.C. Cole's book, *A Hole in the Universe: How Scientists Peered Over the Edge of Emptiness and Found Everything.* This book title allows us an opportunity to do a bit of play on words to introduce one of the central philosophical themes of our thinking. The "hole" in the universe, in our mind, should be restated as the "whole" in the universe. We will support this notion later, but for now let us merely state that Nothing is integral to that wholeness.

There is one more point that needs to be made as we conclude this discussion of the scientific and philosophical fruits of reflecting on Nothing, that which lies beyond the extent, or, as Cole puts it, the understanding that comes from looking over the edge.

Ivo Cohelho, in his book *Hermeneutics and Method* makes an argument that may strike readers as counter-intuitive, yet if they allow themselves to steep in it for a moment, will reside in them as quite true, if astounding. The work of the scientist is a step by step strategy toward a specific end. "But how can means be ordered to an end, when the end is knowledge and the knowledge is not yet acquired?" The answer, according to Cohelho, is what epistemologists—those who study knowledge— call a heuristic, a device that names the unknown suggests some imprecise properties, then utilizes these properties to impose order and guide the inquiry.

Therefore, built integrally into our contemplation of Nothing is a sort of anticipation of what we will learn about Nothing. By inquiring, our mind anticipates the knowledge we are striving for. Bernard Lonergan had great depth on this topic as well. He held that our mind anticipates the knowledge it seeks. This anticipation employs the aid of a heuristic tool. As Lonergan puts it: 'Of themselves, heuristic structures are empty. They anticipate a form that is to be filled."

This means that, as we strive for deeper understanding of Nothing, there is an expectation within us of being able to comprehend Nothing and a general sense of the form of that understanding. Due to this "leaning into" we are motivated to take on the search for Nothing. Also, this leaning enhances the effect of the search as it, inevitably, finds no conclusion in rational knowledge. We tend to fall forward on our faces, and with this falling, we are enriched with new possibilities. By slipping over the edge, by falling off the cliff, we discover how deep the valley actually is.

A Heuristic Tool for Solving the Problem of Nothing

Nihilology, as we have declared before, is indeed a heuristic; it is a powerful way of understanding what a scientist or philosopher can possibly mean when they use the terms nothing, nothingness, no thing, not anything, absence, void, vacuum, and an infinity of others. The heuristic has played a major role in scientific research and philosophy of science debates for well over 400 years. In the early seventeenth century, Joachim Jungius, the wide-thinking German philosopher and scientist, most well-known for his theoretical work on the nature of atoms and a breakthrough conception of the elements, placed heuristic activity at the top of all scientific labor because of its capacity to solve problems that appear unsolvable to plodding thought and even to introduce utterly new paradigms to the science community.

In his article "Philosophical Heuristics and Philosophical Creativity," Alan Hajek gives this helpful illustration of a heuristic. "They say that anyone of average intelligence and moderate talent can become a strong, competition-level chess player by mastering and internalizing certain heuristics. These are captured by slogans such as 'castle early and often' [the only opportunity in chess for a player to advance two pieces in the same move], 'avoid isolated pawns', and so on." These are chess heuristics.

Readers of philosophical and scientific manuscripts are often lost when it comes to interpreting the import of the arguments being put forth regarding Nothing and nothingness. These "over the edge" notions are inherently difficult to stick language to in a valid way. Yet, we all can mature into competition-level interpreters with the help of Nihilology.

We, as readers, can take heart in knowing that some of the blame for our difficulty in understanding discussions in books where Nothing and nothingness come into play lies with the philosophers and scientists themselves who are composing these manuscripts. For, oftentimes, they are not responsibly aware of how loosely they throw about Nihilological language. And thus, philosophers and scientists also could benefit from Nihilology... as a tool for self-criticism.

Just a bit of an aside here. In doing research for this book, we noticed with a smile that there is a very interesting convergence in the etymology of the term "heuristic". The noun derives from the Greek *heuristkein* meaning "to find". Thus, a heuristic is a guide in the speculative enterprise of solving a problem or understanding a conundrum. And the term "eureka" is an exclamation uttered at suddenly breaking through to the solution of a problem or conundrum. It is the from the Greek *heureka,* meaning "I have found." So, a keen mnemonic for this book could be, "Nihilology: a heuristic guiding us to a eureka! on Nothing."

Now, let us return to introducing Nihilology. Hajek gives us this very simple and insightful definition of creativity. "A product—an idea or artifact—is creative to the extent that it is (a) novel and (b) valuable."

Nihilology is indeed novel. The study of Nothing is of growing importance in academic research, both scientific and philosophical and is a ubiquitous buzz in the wider popular culture. Thus, a leading edge, rigorous codification of a methodology for the study of Nothing is an obvious necessity. Yet, until this moment no such leader has been put forth.

At the time of this writing, if one were to perform an Internet search using the keyword "nihilology", only two hits would come up: one a site called Hack Forums whose motto is "Packets Punks and Posts" and the other a site to buy a deck of cards going by the name of Nihilology for playing the game 'Magic: the Gathering." So, it seems that popular culture is at least a small step ahead of science and philosophy in this regard. Also, Nihilology is utterly valuable in that it is a tool for scientists to truly contemplate Nothing. And, as we have said earlier, Nothing is essential to the wholeness of the entire universe.

One more feature of Nihilology: it is a heuristic in the service of Hermeneutics. We need not, at this point, present a sophisticated and exhaustive understanding of the heuristic called Nihilology in the context of the entire realm of the interpretive philosophy of Hermeneutics. But, at least, we need to go beyond the fact that the two terms are perpetually being confused because of that dreaded "eu" in the middle!

At base, Hermeneutics can be defined as the science of interpreting a text. Beyond that, Michael Sellars, in his online article "Heuristics for Good Hermeneutics" helps us out with clear examples. "… a *text* may be a book, an article, something to read—OR something to view, such as a painting or sculpture. In the case of visual texts, the word *read* takes on new meaning. Hermeneutics is most often used by theologians, students of sacred texts, such as the Bible. However, hermeneutics can be applied to any written document. For example, the US Constitution is a piece of literature that requires interpretive skill to determine its meaning and application."

So, Nihilology is tool, a heuristic, that helps a reader solve a problem, for example, what is meant by "Nothing" and how is "Nothing" essential in the interpretation of philosophical and scientific texts.

One final note in the form of a warning: We have seen that "heuristic" was born from a Greek term that means "to uncover." Heuristics help us to observe better, analyze better, interpret better, and apply better our insights into whatever text we choose to study. Yet, we need to be careful. Just like any strictly held ideology, heuristics not only promote insight, but they also, by their very nature, blind us in the way that blinders on a horse limit peripheral vision. While our heuristic Nihilology brings light to formerly dark or obscure arguments of a text, it takes our attention away from other issues in the text. There is risk here, as has been clearly established in the literature of psychology which proves that, when we are faced with uncertainty, we use simple short cuts to help us understand. These short cuts, however, are often more simplistic than simple and lead to biases rather than understanding.

The risk is worth it when it comes to Nihilology, however, for two main reasons. First, most readers and popularizers of science are quite lost in their comprehension of scientific and philosophical discussions where Nothing is germane. Without a heuristic tool, they will not be able to understand these texts at all. Second, many readers come to these texts already filled with biases that have been put upon them by traditional schooling in religious and educational contexts. With the aid of Nihilology, their counter-productive habits of the mind can be overturned. Nihilology is simple, but not simplistic.

(2)
Nihilology as Heuristic

Heuristic/Analytic

Beginning about a century ago, many leaders in the world of science began to voice the view that the great, the truly new, scientific discoveries result from a realm beyond inference, lying outside rational functions. This view reckons scientific discovery to be a unique result of intuition. Max Planck, for example, who received the Nobel Prize in 1918 for his work on quantum theory, stated that if scientists are to be creative they "must have a vivid intuitive imagination, for new ideas are not generated by deduction, but by an artistically creative imagination." Albert Einstein, at about the same time, bravely stated "there is no logical path" that ends at the laws of physics, "only intuition resting on sympathetic understanding of experience, can reach them." Then Hans Reichenbach, author of the 1951 classic *The Rise of Scientific Philosophy* suggested that the act of scientific discovery transcends logic. Any scientist who discovers a theory "cannot name a means by which he found the theory," only experiencing "that he saw intuitively which would fit the facts."

This upvaluation of the role of intuition in the discoveries of science and philosophy of science is a recent one in terms of the long and esteemed history of these disciplines. And there are still many today in the scientific community who disagree as a deep matter of principle. They hold that there is no role for intuition in science. For example, there are these words, rather snippy for a rationalist scientist, from Kyle Hill in his web article titled "Thinking Scientifically." "By intuition we mean vague feelings or gut reactions about a question or phenomena. One problem with intuition as a knowledge source is that our intuitions are often wrong… Mood, and a host of other psychological and physiological factors influence intuition. A judgment of information may be solely based on how you rolled out of bed this morning."

We hold neither of these views. First of all, there is no evidence that all important scientific discoveries result from intuition. For example, Richard Feynman, the creator of the famous Feynman Diagrams that illustrate the movements of sub-atomic particles, was known among his colleagues as the "Great Calculator." As his long-time colleague Stephen Wolfram tells it in "A Short Talk About Richard Feynman," he would start by filling up page after page with handwritten calculations. Having found a solution, Feynman would rework the process in order to synthesize a seemingly simple explanation. Feynman would play his showman's game by striking awe in his students with his seemingly effortless magic mind, never revealing his pages of calculations stored away in his briefcase!

So, we must accept that great discoveries, like the Feynman Diagrams, lie at the end of many pages of mathematics, with no gap in rationality of any kind. On the other hand, creative science ***is*** often brought to bloom through intuition. And the heuristic questions of Nihilology are sharp cuts that can aid in both rational and intuitive solutions to problems of Nothing and nothingness.

We do not use the term "short cut" because it connotes that one is lazy and so employs a quick way to solve a problem. We hold that a heuristic allows entry into a problem that is unsolvable without it. Imagine you have a hot bowl of miso soup in front of you and you are famished. You need to get the vegetables and tofu into your stomach as soon as possible, not being able to wait until the broth cools. You pick up a pair of chopsticks and dig into the bowl. Ah, food! These chopsticks are not a shortcut, one that misses the subtler aspects of the soup. The chopsticks allow you to eat the miso soup at all!

We also do not hold that insight has anything to do with vague feelings or gut reactions, mainly because this view ignores the overwhelming narrative evidence concerning scientific discovery. And insight certainly has little to do with how one rolled out of bed in the morning!

Now, the process by which James Watson and Francis Crick discovered, in 1953, the double helix "twisted-ladder" structure of DNA narrates for us an excellent example of how insight actually works in scientific discovery. Their discovery has been called a milestone in the history of science as it guided scientists in deciphering the genetic code. The image of the double helix has even worked its way into our cultural mythos: it has been sculpted, painted, worked into brand logos, bent into jewelry, and even extruded into plastic toys.

When these two young men met in 1951, they immediately took on a zealous, eighteen-month-long, absolutely focused search for a three-dimensional model for DNA. They studied genetics, biochemistry, chemistry, physical chemistry, and X-ray crystallography. As stated in a 1953 article in Nature Magazine, "Relying on their brilliant intuition, persistence, and luck, the two showed that DNA had a structure sufficiently complex and yet elegantly simple enough to be the master molecule of life."

A female (competitor?) colleague of the two named Rosalind Franklin at King's College in London had succeeded in shooting clear X-ray images of DNA fibers that suggested a helical shape. With this image as their heuristic guide, the time grew ripe for their big discovery. Watson was seriously playing with accurate cardboard cutouts of the relevant molecules when he was hit by a bolt of genius: **A** conjoined with **T** echoed a combination of **C** and **G**. These dyads fit nicely between the two helical sugar-phosphate backbones of DNA, and these backbones ran in opposite direction to each other, one up, one down.

The two scientists were obviously overwhelmed with their insight. Crick trumpeted to all the pint-drinking patrons at the Eagle pub in Cambridge, "We have found the secret of life." Thus, following a long contemplation on a focused problem, guided by a heuristic question ("How can we create a three-dimensional model?") then by heuristic images (X-rays of DNA's helical shape in conjunction with cardboard cutouts) the search finally culminated in a flash of insight. Thus, we say that the research process of Watson and Crick is a perfect case study in the use of heuristics leading to intuition in science.

This energized thought released at the eureka! moment is attested to by Arthur Koestler in his now classic work from the 1960s, *The Act of Creation.* "The eureka! cry is the explosion of energies which must find an outlet since the purpose for which they have been mobilized no longer exists; the cathartic reaction is an inward unfolding of a sort of oceanic feeling." The exhilaration of the eureka! experience can show itself in understandable hyperbole. "We have found the secret of life."

The Scissors of Nihilology

To conclude this discussion then, let us state that, over this past decade of our investigations into Nothing, the confidence has solidified in us that heuristic reasoning, often culminating in intuition, and analytic reasoning, often regarding the products of intuition, are *both* essential to a trustworthy scientific method. We advocate for what, in the literature of the psychology of thinking, is termed dual process cognition. Let us once again call upon Ivo Cohelho as he illuminates the thought of Bernard Lonergan, this time regarding the mathematical research process, "[Mathematics] sets up a scissors movement... Not only is there a lower blade that rises through data through measurements and curve fitting to formulae, but also there is an upper blade that moves downward from differential and operator equations and from postulates of variance and equivalence."

Though from a field slightly far from our discussion, this upper blade and lower blade of discovery and empiricism provide a very useful image that we shall borrow from Mr. Cohelho and apply to the working of heuristic thinking (sharp cuts) in coordination with analytical reasoning to cut through a problem in the interpretation of texts on Nothing. They are the scissor blades of Nihilology.

The noted astrophysicist from Cornell, Carl Sagan, who had "billions and billions" of followers in his career, spoke out very clearly in his foundational book *Cosmos* regarding this need for heuristics and analysis. "[Valid cosmic discoveries] require skepticism and imagination both. Imagination will often carry us to worlds that never were. But without it, we go nowhere. Skepticism enables us to distinguish fancy from fact, to test our speculations." Emiliano Ippoliti, in the editor's essay to the book *Heuristic Reasoning,* does a more technical take on Sagan's words. He introduces the terms "ampliative reasoning" which encompasses heuristics as well as other imaginative techniques and "non-ampliative reasoning" which encompasses deduction and other methods for corroborating imaginative discoveries. In the end, he concludes good science requires both ampliative and non-ampliative reasoning in the evolution of knowledge.

Yes, Nihilology leads scientists and philosophers to solve difficult problems, perhaps even to the intuitive eureka! It is a heuristic tool to penetrate a problem of interpreting Nothing. However, analytical thinking is also necessary for a complete understanding of a problem at hand. Nihilology is rigorous in the realm of clear definitions of terms, grounding analysis, and extensive research. Though we indeed hold that Nihilology is a sharp cut heuristic, it is also an analytical tool as well.

One last note. When Rudy Rucker, the well-known American theoretical mathematician, had his first meeting with Kurt Gödel, he was a bit intimidated. Gödel, who we discussed before, began the set theory revolution with his Incompleteness Theorems. He was rumored to be a rather strict European who held his own in intellectual boxing matches with Einstein during their walks about Princeton. Yet, when they met, Rucker beheld "a joyful twinkly sage—not some obsessed fossil. What struck me most about Gödel was his intellectual freedom—his ability to move back and forth between frankly mystical insights and utterly precise logical derivations."

Gödel's comfort with sweeping insights and with logical analysis is certainly the ideal researcher's skill: balance and comfort with heuristically guided insight as well as rigorous analytical probing. This is a skill to which all of us should aspire, in our personal pursuits and in professional studies of philosophy and science.

Image as Trans-Heuristic

In the opening paragraph of this book, we mentioned a paper Gustavo had written while a graduate student at the University of Toronto that gave birth to our decade-long fascination with Nothing. During that same period of study, he also had a seminal intellectual experience that lent depth to our search for Nothing.

As part of a seminar on the history of science, Gustavo's professor began by explaining the Plum Pudding Model of the atom that had been proposed at the dawn of the twentieth century by the esteemed Nobel Prize winner from England: Sir Joseph John Thomson. (Other scholars poked fun at Thomson by naming his model after a favorite English Christmas dessert. I have a feeling he never quite enjoyed Christmas dinner again!) Thomson, upon discovering electrons, which he gave the delightful name "corpuscles," proposed that these negative charges moved randomly about in a positively charged ooze that his colleagues likened to pudding.

The professor then projected this image of what has come to be known as the Rutherford gold foil test that took place about five years after the Plum Pudding Model. In this experiment, Alpha particles from an emitter were aimed at a sheet of foil. The results were indeed startling: almost all particles flew straight through, yet a few reflected immediately back at the emitter. Rutherford concluded that there is a small positive charge at the center of the atom where almost all of its mass is concentrated. We now call this mass the nucleus of the atom. There is no pudding of positive charge around randomly dispersed negative charge particles. In fact, the vast region of the atom is essentially empty. (It is often said that Rutherford was afraid to get out of bed the next morning for fear of falling straight through the floor!)

Now this gold foil diagram opened Gustavo's eyes immediately. He excitedly exclaimed to the professor, "This is the first model of the void!" Rutherford's intent was to depict a model of matter, but Gustavo had an insight into the void. And this was the moment that intensified our scientific search for Nothing. Indeed, he had seen the potency of images as heuristics toward the solution of a scientific conundrum.

Furthermore, in doing research over the past decade, we repeatedly found that many of the breakthrough scientists found their way to insight using heuristics not involving words or even thoughts *per se*. They often used pictures, geometric diagrams, or three-dimensional models. Our sense is that pictures and images are trans-heuristics unto themselves. A bridge that leads to the cliff edge of rationality. This understanding of images as heuristics in scientific discovery makes Nihilology unique in the field of science writing.

We've called the discovery of the double helix structure of DNA by scientists James Watson and Francis Crick a perfect example of scientific insight. Now, let us explore farther just how that double helix insight came about.

In 1951, Watson and Crick had just begun to work together in a Cambridge laboratory on the DNA problem. They were an odd pair of colleagues. Watson was a twenty-something American with a fresh Ph.D. Crick was a thirty-something British graduate student. Neither were chemists and neither had built a prestigious career that would bolster their being heard among academic circles. Yet their intense drive to penetrate the DNA structure and their quirky creative processes led to a Nobel Prize-winning breakthrough.

In mid-1953, our pair read a paper by the eminent American molecular biologist Linus Pauling wherein he proposed a structure of DNA as a three-stranded helix. They immediately concluded that Pauling had made a critical mistake, yet they also knew he would soon uncover his error. In some unknown mathematics of prediction, they estimated that Pauling was six weeks away from the correct model. By that time, Pauling was already a supernova in science. Excited and driven by the career rewards of beating Pauling to the goal, Watson and Crick lit into a four-week frenzy, a month-long non-stop experimental contemplation that resulted in their eureka!

If we look closely at the writings of these discoverers themselves, we see that they unashamedly hold that their historic insight came about not through rational investigation alone, but through the essential dynamics of aesthetic beauty, playfulness, and model building.

Crick and Watson both published extremely popular (and by science publishing standards very slim) books revolving around their famous discovery. Watson's account *The Double Helix: A Personal Account of the Discovery of the Structure of DNA* came out in 1968. Crick followed in 1988 with his *What Mad Pursuit: A Personal View of Scientific Discovery.* It is revealing to note that both scientists chose to use the word 'personal' in their subtitles. For, indeed, safely separated from the original discovery by fifteen and thirty-five years, with a Nobel Prize in hand and world fame to give them license to softly cross the barrier of objectivity in science writing, they could reveal the workings of scientific progress from a fleshier point of view.

Crick discloses early in his book a long-withheld disappointment in the rational ways of science by pointing out that a winding history of theoretical work in attempt to break the genetic code had been an utter failure. Later in the work, he quite clearly states that Watson came to the ultimate solution to the puzzle of the helix of DNA "not by logic but by serendipity." Note that Crick opposes logic to serendipity, thinking as opposed to luck, i.e., some unknown mechanism that is not thinking. We will not let this unknown go quite so easily. In the next section titled "Unconscious Scientific Discovery" we begin an analysis of the mechanics of this "not thinking."

Crick also embodies in the book title the non-rational nature of his and Watson's scientific endeavor. Madness, as he implied, is ironically the pathway to discovery in science. To go one step farther with this title, we know that the phrase "mad pursuit" is borrowed from the poem *Ode on a Grecian Urn* by John Keats, an early nineteenth century English Romantic poet to whom Crick often referred. The phrase in the poem connoted a sexual ecstatic chase of virgins by excited hormonal young men. Thus, for Crick, scientific discovery is akin to an aroused endeavor of the body and emotion and beauty rather than simply a rational affair.

We should be mindful also that *Ode on a Grecian Urn* contains this famous verse in its culminating stanzas, "Beauty is truth, truth beauty." This is no coincidence; it seems quite relevant to our discussion of Crick's view of scientific innovation. For example, in one section of his book, he quotes the English painter Francis John Minton, a large character in the London art scene in the 1950s, who said of his own artistic creations, "The important thing is to be there when the picture is painted." In using this reference, Crick implies that the creative moment, the eureka! of DNA and by extension all scientific breakthroughs, are a matter of witnessing an event that comes about by force of some not quite intellectually-initiated process. Ultimately, his double helix was experienced as an aesthetic creature, not a rational discovery. "It's so beautiful you see, so beautiful."

Now, Watson, in his book, reveals an even more central and pervasive role for the aesthetic in science. He repeatedly employs the notion of beauty as a guiding principle. Indeed, if Watson had a pet word to connote the driving force of their mad pursuit, it was "pretty." As he states in one of his ebullient propositions, "Perhaps the whole problem would fall out just by our concentrating on the prettiest way for a polynucleotide chain [DNA] to fold up."

Watson even goes so far as to hold that the aesthetics of a scientific solution serve as probative for the solution itself! He notes that many scientists initially criticized their conclusion, holding that, although their model of the DNA double helix was aesthetically pleasing, this shape for a sugar-phosphate backbone simply could not exist in fact. Watson looks back with glee, "Happily now, we knew that this was not true, so we had lunch telling each other that this structure was just so pretty that it had to exist."

To go one step farther, it is a well-known fact, even in our popular culture, that Watson found the eureka! of the double helix while playing around with various configurations of base pairs, using three-dimensional cardboard cut-out models. (This is a bit of a pop simplification in that, at various junctures, certain components of the models were wire and metal shapes cut out by Watson or by the lab's machine shop.)

Crick states that Watson's essential discovery was how exactly the base pairs or nucleotides fit together, that is, how adenine-thymine and guanine-cytosine connect the strands of DNA to form the helix shape. He then states that this momentous discovery in the history of science was mostly a matter of luck, as are, according to Crick, most breakthrough insights in science. Then Crick goes on to add one more non-rational ingredient to the mix leading to insight. "This episode emphasizes that play is often important in research."

It wasn't just that Watson was driven by the beauty of the DNA helical structure; it was also that he played with cutout models, played toward a very serious end. We will soon discuss the foundational work by Graham Wallas *The Art of Thought*. Yet it should be mentioned here that Wallas, regarding the search for insight, was an advocate of the need for a playful mind that he describes as one that juggles images and ideas, imagines hypothetical scenarios, and even entertains situations that are rationally held to be untrue. Watson himself relates proudly that his main tools were a simple array of model molecules that reminded him of toys he had as a preschool child. "All we had to do was construct a set of molecular models and begin to play, and with luck the structure would be a helix."

As an interlude at this point we'll mention two parallel aspects of the mad pursuit toward the double helix. Watson describes the pendulum quality of his search in the final days leading to his playful insight. After working in his office all day he spend almost every evening at the movies, always hosting in the recesses of his mind a quiet dream that at any moment the answer would bloom and they would have a stereochemically valid configuration for the DNA backbone.

This rhythm of intense conscious work followed by periods of rest foreshadows the analysis below contained in Graham Wallas's four stages of insight. Going to the movies accords with stage 2 that Wallas terms Incubation, a period away from intensive conscious investigation of a scientific problem where one rests, puts one attention into irrelevant matters, and allows the unconscious to take over the pursuit.

Watson also makes a second interesting observation about his mental and aesthetic state on the day after their eureka!, "The following morning I felt marvelously alive when I awoke... staring up at the pinnacles of King's College Chapel that stood out sharply against the spring sky… thinking that much of our success was due to the long uneventful periods when we walked among the colleges…"

Watson seems to have been enjoying what Romain Rolland, the early twentieth-century philosopher, termed the "oceanic feeling." This is a state of psychic-float that occurs when the driving force of an intense scientific pursuit is released at the eureka! of solution. Rolland held that this experience lay at the basis of all religions. However, we must remind ourselves, that neither Watson not Crick were religious people in the sense of American Protestantism or the Church of England. The mythos of these traditions was not at all compelling to them.

Now, let us return to the final aspect of Crick and Watson's discovery: the use of the three-dimensional images afforded by models. Earlier we argued that employing images in this way constitutes a trans-heuristic bridge from the realm of rational thought to the realm of insight.

It is so important to emphasize that Crick and Watson never did any actual experiments on DNA. They read and learned from the x-ray crystallography performed by Rosalind Franklin from her time at King's College London. They followed the findings of electron microscopy by Robley Williams at the California Institute of Technology. Yet they themselves laid all their faith in building models as the way to the discovery. They were confident that the structure of DNA was more amenable to three-dimensional visual understanding. Thus they utilized the trans-heuristics of images to find their eureka!

One additional means of gaining perspective on the DNA discovery process of Crick and Watson comes from the findings of brain science. This research tells us that aesthetics, play, and three-dimensional models all exhibit movement from the left hemisphere of the brain to the right hemisphere. Movement to the right brain is away from conscious rationality and language toward more visceral, emotive, non-rational experience of the world. Crick and Watson were experts at right-brain scientific creativity.

Ian McGilchrist, in his book *The Master and His Emissary: The Divided Brain and the Making of the Western World,* outlines the basics of the left/right hemisphere functions. The human brain has two regions, called hemispheres, and each hemisphere attends to the world in a different way, yet "the ways are consistent" with one another. The right hemisphere sees things as "whole" and in their context, whereas the left hemisphere sees things abstracted from context and broken into parts.

For our discussion, it is revealing as a valid extension of his analysis, to substitute McGilchrist's term "whole" with the term "solution" to a scientific problem. Thus, a left-brain solution is experienced as a solution by the left brain and a right brain solution is experienced as a solution by the right brain. These solutions are of a very different nature, yet "the ways are consistent." Hence, as we argued before regarding analytic and heuristic thinking, the solutions to scientific problems require both hemispheric functions: the language-based reasoning of left and the aesthetic play of spatial images of the right.

Watson and Crick, in the end, were not only believers in their right-hemisphere model play. They never gave up the balance offered by left-hemisphere thinking. For example, they had a collegial debate strategy that they employed in a very athletic manner while in their offices at the laboratory. One would bring up his latest theory regarding DNA and the other, good theory or bad, would do his best to destroy it with reasoned critique. These matches of the mind were so loud and on-going that their colleagues conspired to move them into the same office at a distance from the rest so their din would not bother others' work!

Another illustrative example of an image as a trans-heuristic in scientific discovery is the Penrose Diagram that depicts in a two-dimensional image the utterly complex relations of different points in space-time... which is already an abstracted mathematical model that unifies infinite space and time. Helene Mialet, the well-regarded anthropologist of science wrote a book titled *Hawking Incorporated: Stephan Hawking and the Anthropology of Knowing* that is very helpful in simplifying such complex notions. She explains that the Penrose Diagram is not really an answer to a problem, rather a heuristic tool, a spatial visual that can tweak intuition when one gains a facility to see them from a particular angle. In the end, Penrose Diagrams, according to Mialet, are a sort of prosthetic device!

Penrose Diagrams have acted as ground breakers in the discovery of the structure of black holes, the existence of multiple universes, and the positing of countless space-times. They provide us with a window on the Olympic-level gymnastic pictures used by physicists to gain insight into questions that are too complicated to work out exactly through mathematical calculations. (Steven Hawking was a great fan of the Penrose Diagram, as we shall see below.)

Support for our analysis of images as scientific heuristics comes from fields outside science as well. Professor Julia Marshall, noted art educator, wrote an overview article titled "Image as Insight: Visual Images in Practice-Based Research" that has greatly enhanced our understanding of images as heuristics in scientific discovery. Her breakthrough notion is that problem solving and meaning bloom when one morphs ideas into visual images or other visual experiences. This metamorphosis allows information to be **seen**, to be experienced in a visual way which gives forth new insights. Thus, image creation is primarily an exercise in knowledge construction: a process of coming to know. Marshall goes one step further to suggest that the sort of knowing created by images, in contrast to rational thinking, is much more wholistic and synthetic in nature.

So we take from this that, when scientists are working hard on a tough problem, they can—and often do according to history—transform their thoughts, calculations, and words into images. This transformation, this very act of creating an image, is a new form of knowing. And, in turn, those images allow for the possibility of new insight into the problem at hand.

Throughout this article, Marshall argues impressively that all these bold notions about art creation apply to the scientific use of images as well. We had independently been thinking along the same lines and were thus gratified to have her support in this interpretive quest. Together we conclude that images serve as trans-heuristics that lead the scientist, and artist, to the bounds of rationality, to the eureka! of insight, and to new, more global understanding.

Support for our contentions about images as heuristics comes not only from the world of art research, but also from theology. Bernard Lonergan SJ, one of the best and most erudite of thinkers in this area, wrote, in a quirky yet profound way, about how to establish the definition of a circle based on the image of a cart wheel. He avers that if one rids the mind's eye of the image of the cart wheel's center, spokes and curve then one loses their grasp of roundness, "But it is that grasp that constitutes the insight... The pivot between images and concepts is the insight."

One further aid in our understanding of the heuristic nature of images has been the book by Arthur I. Miller titled *Insights of Genius: Imagery and Creativity in Science and Art.* Miller dedicates much of this work to exploring how what he calls "imaginative insight," or insight stimulated by images, has shaped the great breakthroughs in twentieth-century physics.

Miller, who holds a Ph.D. in physics and is an emeritus professor in History of Science, contends, as we do, that images are utterly critical to new understanding in the search for solutions to scientific problems. For one reason, they give embodiment to what would otherwise be a random and confusing disarray of rational information. Further—and this is so important to our later critique of the popularizers of physics—they also provide the "escape velocity" for the mind to break free of the gravity of logic in order to reveal new insights.

Here is a perfect example from Miller's sweeping overview of how images led to historic new insights in science. Beginning about 1923, with the postulation of Bohr's theory and Heisenberg giving birth to quantum mechanics, there was essentially no use of image in mainstream physics. Vision had been lost! Miller notes, "The term 'representation' was taken to be synonymous with mathematical formalism..."

This all changed radically in 1948 with the introduction of the famous Feynman Diagram. Richard Feynman was a rock star in the world of physics whose pictorial representations of the mathematics describing the behavior of subatomic particles allowed physicists to actually carry out extremely complex calculations visually. They offered a **glimpse** beyond the findings of Galilean-Newtonian and relativity physics. In 1950 Werner Heisenberg deemed Feynman Diagrams to be the "intuitive [*anschaulich*] contents" of the new theorizing called quantum mechanics.

Now, there could be no more apt scientist than world-wide idol Steven Hawking as the epitome of expertise in the use of heuristic images in the process of discovery. It seems that Hawking gave up on sophisticated mathematical calculations in the midst of his career due to his disability that stole almost all his physical capabilities. In Hawking's own words from an in-person interview with Helene Mialet:

> **Hawking**: "I regard my disability as a minor detail like being color blind. If you are colorblind you develop tricks to recognize traffic lights and clothes.
>
> **Mialet**: "What are the tricks for you?"
>
> **Hawking**: "Pictures."
>
> **Mialet**: "In your mind?"
>
> **Hawking**: "Yes."

Kitty Ferguson, the well-known science writer, in her book *Steven Hawking: An Unfettered Mind*, attests to the astounding ability Hawking had developed: "[Steven shows] an unusual ability to manipulate mental images of objects, curves, surfaces, shapes, not merely in three dimensions, but in the four dimensions of space time." Hawking himself said that he can't actually envision four dimensions. Rather he saw an internal image of two dimensions at a time while retaining a sense that they are part of a four-dimensional whole. This is in itself an astounding feat of visualization.

Mialet concludes with this perspective: Hawking transformed his handicap into an advantage. He created a work-around for his muscle-based limitations by redefining questions so he could see in his mind's eye either their solutions or the frameworks in which problems were posed. In short, he solves problems by manipulating images, often in the form of sophisticated models, rather than equations.

As depicted in *The Theory of Everything*. the wonderful Academy Award-winning movie about Hawking's life, he is well-known for having moments without thought, without digitally created words, without "cheek-generated speech" that result in insightful breakthroughs. He is indeed a master of the image-based heuristic.

Unconscious Scientific Discovery

Popularizer scientists may argue that there is no value or need to analyze such notions as images providing "escape velocity" for investigators to break free of the realm of concept because, at base, there can be no empirical proof of the reality of a state beyond concept. This is, indeed, a superficial view and one that has been undermined by a long history of investigation into the nature of scientific insight.

In 1926, the British social psychologist Graham Wallas introduced his four-staged model of creative thinking in the book *The Art of Thought*. His model is still the standard format for researchers in the area of creativity. The first stage he calls Preparation, a dedicated application of rational thought that is difficult, conscious, and without a rewarding solution. This is the stage that describes the "four-week frenzy" of Watson and Crick, preparing them for the DNA discovery.

Wallas's stage two is Incubation (expounded below). Stage three is Illumination. This is the eureka! moment of insight. Wallas employs his delightful turn-of-the-century sensibility in calling this eureka! "the happy thought." Finally, there comes stage four: Verification. This is the stage where an insightful scientist returns to the lab or the telescope and attempts to download their eureka! into mathematics, measurement, and new hypotheses.

The Preparation stage describes the classic situation for a scientist who has been working diligently on a problem for a long time without success. Gary Klein, in his excellent book *Seeing What Others Don't: The Remarkable Ways We Gain Insights,* suggests that such a scientist enters into a psycho-emotional state he calls creative desperation, a fertile intellectual condition that reminds one of the play by Jean-Paul Sartre, *No Exit.* As Klein says, this condition "requires finding a way out of a trap that seems inescapable."

The notion of creative force being born out of a trapped intellect is not simply one of the modern writers on psychology and science. Anselm of Canterbury, who seems to keep coming up in our discussions, gave witness to this force in his *Proslogion* way back in the eleventh century. Please forgive this longish quote, but his language is true archaic beauty:

> ...I began to ask myself whether perhaps a single consideration could be found
>
> which would require nothing other than itself for proving itself
>
> and which would suffice by itself to demonstrate
>
> (1) that God truly [i.e., really] exists and (2) that He is the Supreme Good
>
> I often and eagerly directed my thinking to this [goal]
>
> at times what I was in quest of seemed to me to be apprehensible
>
> at times it completely eluded the acute gaze of my mind
>
> At last, despairing, I wanted to desist
>
> as though from pursuit of a thing which was not possible to be found
>
> But just when I wanted completely to exclude from myself this thinking
>
> just then it began more and more to force itself insistently upon me
>
> unwilling and resisting [as I was]

Then one day when I was tired as a result of vigorously resisting its entreaties
what I had despaired of [finding] appeared in my strife torn mind
in such way that I eagerly embraced the [line-of-] thinking
which I, as one who was anxious, had been warding off

This is an excellent example of advancing through Wallas's first three stages of creativity. Anselm thought and thought "often and eagerly" unto exhaustion and "despairing" (Preparation). He then entered a period when he "wanted to desist," to put distance and barrier between his attention and the futile pursuit. He attempted to "exclude from myself this thinking," to "vigorously resist its entreaties" (Incubation). Finally, he awakened to stage three Illumination, his most famous insight and a staple ever since of theological proofs for the existence of the personal divine: God is that greater than which nothing can be thought.

It is revealing to note that the insight "forced itself insistently upon" Anselm. The insight came from a realm not experienced as his own mind. Yet, where is that place beyond Anselm's mind? Answering this question has led us on a fascinating research path that has culminated in our own insight. We will, in fact and with solid argumentation, answer the question. It plays a central role in our understanding of Nothing and the development of our heuristic Nihilology. It will add to our tools, and confidence, in critiquing the popularizer scientists.

In his classic work *The Unconscious Before Freud,* Lancelot Law Whyte outlined a history of the notion of the unconscious mind. He posited that, in the West, a momentous cultural shift occurred in the seventeenth century as witnessed in the philosophical dualism of René Descartes. Human beings made a choice to believe that the defining characteristic of mind is awareness. This choice gave birth in the Western culture to the notion of an unconscious mind. And the search for the seat and nature of unconscious psychic functioning began in earnest.

It was, of course, the work of Sigmund Freud in the early twentieth century that deepened our understanding—and popularization—of the unconscious mind. It is Freud's unconscious that has dominated in the West until just recently (with the new appreciation of trans-personal psychologies and the importation of eastern religions.) He laid out a sort of interior typology that envisioned the conscious mind as the tip of a profound psychic iceberg. The overwhelming bulk of the psyche was constituted of the unconscious that, according to Freud, steers and manipulates a great deal of our conscious emotions and behaviors.

Now, Freud was not the creator of the notion of an unconscious within the psyche, yet he held a decidedly characteristic perspective on what lurks there: primitive, sexual, atavistic forces that need to be resisted from appearing in consciousness. In his speeches, Freud gave dramatic descriptions of the conflicts that rage in his patients between ego (conscious) needs to repress and id (unconscious) demands to express. They pushed pathogenic experiences out of consciousness, ones that "involved the emergence of a wishful impulse which was in sharp contrast to the subject's other wishes and which proved incompatible with the ethical and aesthetic standards of his personality." Freud saw the dynamics of the psyche as a conflict of opposing forces, a brutal struggle between the two main contingents of the mind.

Graham Wallas took up the Freudian psychic topology and bore its standard regarding this one essential tenet of Freud's teachings: the psyche can register interactions with the internal and external world without having any conscious experience of doing so. These registered events can, according to Wallas, be brought into consciousness via the techniques of psychoanalysis such as hypnotism, free association, and dream analysis.

For most of the past century, much of the search for the nature of the unconscious has taken the well-reputed entranceway of psychoanalysis. Yet in recent decades, with the sophistication of neurological measuring tools, a new means of entry has opened up. As neuropsychologist Rhawn Joseph states in his book *The Right Brain and the Unconscious: Discovering the Stranger Within*, "the so-called unconscious mind is a function of right brain and limbic system mental activity."

Joseph shows that there seems to be evidence that the unconscious has right brain activity correlates. We could say that Freud located the unconscious deep within the human head and neuroscience has found it residing in the right hemisphere of the brain.

Now, much of the neurological scientific research into the unconscious has been focused on the act of seeing. It has been well-established that human beings, among animals, are highly dependent on and extremely sophisticated in our sense of vision. Thus, it is held among experimenters, that seeing is a fertile realm of research toward understanding human consciousness and unconsciousness. (It is interesting to note that Francis Crick, after his discovery of the structure of DNA, devoted much of his later career to the investigation of vision as a means to discover the human essence. In 1994, he published a major book titled *The Astonishing Hypothesis: The Scientific Search for the Soul.*)

In 1976 Julian Jaynes published his book *The Origin of Consciousness in the Breakdown of the Bicameral Mind* that made popular the entire analytical language of "left brain vs. right brain." In it Jaynes analyzes the structures and processes of human vision. He points to the fact that if we have a small image of an object on our retina, we automatically see the object as located in the distance; we are not conscious of making any sort of correction. He elaborates with other examples of what are termed perceptual constancies: how we manipulate color, light, and contrast—in our waking and dreaming lives—without our being at all conscious of what is occurring within us.

The work of Julia Marshall is once again very perceptive. She points out that most scholars who study seeing conclude that it is a multi-faceted process that simply cannot be exhausted by any one reductive assumption. Visual imagery, as a human modality, takes place with and without human consciousness, and images can stimulate and articulate ideas in ways that words cannot.

This line of argument harkens back to our discussion of how visual images work as trans-heuristics toward scientific insight. Yet, in her main argument here, Marshall makes it clear that, based on the research into vision, human beings are not conscious of and have no actual experience of many of the most intimate processes of our own minds.

Furthermore, back in the seventeenth century, when making the choice of awareness as the true dominion of our mind, we gave birth to puffed-up pride in the dominance of the conscious mind that most researchers today agree is a false one. This false pride in consciousness will come to play below in our critique of popularizer scientists. As Julian Jaynes chides us, consciousness is a much meeker function of human mentality than we unreflectively believe, basically because we are not conscious of that of which we are not conscious! "It is like asking a flashlight in a dark room to search around for something that does not have any light shining upon it. The flashlight, since there is light in whatever direction it turns, would have to conclude that there is light everywhere."

So, if it is factual, as wide-ranging research has concluded, that psychic activity takes place as a common and necessary process within us—without any awareness—what does this say of our psychic goings-on regarding scientific insight? Well, it seems that, in contrast to the routine problem solving, significant breakthroughs in scientific research often are born outside of consciousness, outside of deliberation. As Gary Klein concludes, "They come without warning. Our minds do our heavy lifting without our awareness."

Now Klein came to this culminating position in 2013, yet it was held to as early as the eighteenth century in the person of the great "critique" philosopher, Immanuel Kant. He proclaimed that genius (his term for insight) cannot be explained scientifically as to how it brings its products about. In his wonderful baroque language he says, "That is why, if an author owes a product to genius, he himself does not know how he came by the ideas for it; nor is it in his power to devise such products at his pleasure or by following a plan..."

Some researchers have even assaulted our pride in consciousness to such an extent that they relegate certain streams of reasoning itself to the unconscious. Julian Jaynes leads this assault in holding that consciousness is, indeed, not the sole seat of reason and—yet a deeper cut—maintains that the most difficult challenges of creative reason are undergone without any attending consciousness at all!

This challenging notion of unconscious reasoning is not only a contemporary proposition. L.L. Whyte held it back in the fifties and he harkens way back to the philosophy of mind as taught by the great seventeenth century Gottfried Leibniz whom today's cognitive scientists still call on for guidance. According to Whyte, consciousness is not inherent to ideas. Ideas with consciousness he calls apperceptions, à la Leibniz, whereas ideas without consciousness he calls perceptions, or "dark images." Ideas with consciousness often come to be as a product of ideas without consciousness.

Arthur Koestler offers up the final deprecation of our false pride in conscious thought. In the end, he says, it is no different than the childlike act of riding a bike. Everybody can ride down the street, but nobody knows how it is done! Bike riding is governed by implicit codes, yet engineers and bicycle manufacturers have tried in vain to come up with the formula. "In other words, there is less difference between the routines of thinking and bicycle-riding than our self-esteem would make us believe."

In the modern world of cognitive research, it is widely concluded that problem solving goes on in both realms, conscious and unconscious. (Yes, Freud's unconscious as the realm reserved for dark atavisms in need of repression has indeed been eclipsed.) We mentioned before the research of Arthur I. Miller that helped in our understanding of images as trans-heuristics. He states that scientific insight is the child of aware intellectual thinking in partnership with "parallel unconscious processing of information." He points to the nineteenth century German thinker often tagged the Father of Modern Psychology, Wilhelm Wundt, who bravely proposed that the human mind receives ideas as gifts from unknown psychic regions via unknown psychic processes. "This unconscious mind is like an unknown being who creates and produces for us, and finally throws the ripe fruits in our lap."

This picture painted by Wundt is a very fertile one for our discussion of the nature of scientific insight. First it fixes in our mind's eye the fact, as we have elucidated in this discussion, that unconscious processes are essential to the eureka! moment. We see a scientist sitting cross-legged at the base of an apple tree in autumn, resting their mind from a grueling investigation of a vexing problem. Suddenly, and with great surprise, a ripe apple, fully polished, falls from the tree into their lap. They eat of the apple and it is sweet.

What Wundt's painting does not depict fully is the apple tree itself. Our scientist is seen with their back against the trunk. They may have some peripheral sensation in their spine as to its presence. The few brush strokes devoted to the trunk hint at there being something up, something above, but the branches and leaves and ripening fruits of the tree exist beyond the picture frame.

"Beyond the picture frame" is akin to our statement that scientific insight of necessity involves the unconscious psyche. However, we cannot rest with this proposition, because it says very little more than this: insight comes from the "un," from a place that we don't know much about. A proposition with equivalent value to the few brush strokes for the trunk in Wundt's painting.

Nonetheless, the authors, scientists, and science writers we have presented in this section do indeed rest with the "un"—the place where insight is born. Graham Wallas called this a place of "incubation," wherein the happy thought somehow stays warm and sprouts. Anselm of Canterbury said it was place where he "excluded [him]self from thinking." Julian Jaynes likened it to a dark room "that does not have any light shining upon it." Arthur Koestler holds that this place is "governed by implicit codes," and our painter Wilhelm Wundt calls it "an unknown being."

These various "un" propositions were often bold unto themselves in their day and in the context of the arguments the writers were putting forth. Yet, for our development of Nihilology, they are only a big first step. We need to, and will indeed, look explicitly into this dark place. In a section below called "The Nothing of Religions," we will begin. In the section thereafter, titled "Nihilitive Knowing", we will present our culminating conclusions. Just a hint of foreshadowing: this place is not so dark; it is the realm of Nothing.

Let us conclude here by stating that it is essential for our discussion of Nihilology to accurately describe the psychological process of insight, one that explains how the eureka! comes about in science. This line of thought will aid us in critiquing the popularizer scientists who espouse the reality of intuition yet offer no argument about what exactly the process of intuition entails.

From our research thus far, it is our understanding that a focused rational contemplation on a scientific problem is aided by heuristics. The work of intense thinking causes rationality to fatigue and go quiet for short periods of time, providing a restful gap in the reasoning process. Images often come into play as a trans-heuristic bridge from the realm of the conscious thought to the realm of rest where unconscious parallel processing of the problem can occur. When one returns to thinking from the gap, the first thought will be energized by the rest of the gap. Eureka!

Our model is new; it finds a more scientifically satisfying explanation of the source of the mental elation released at the eureka! moment than the classic four-staged model put forth by Wallas. It also fully accounts for the processes occurring within his shadowy Incubation stage. These evolutions beyond Wallas will come to full fruition in the next three sections of our book. Yet, for now, let us state that intense rational thought has been shown to be, by its very nature, a mental energy drain. The eureka! elation is a product of the period of rest when thought is transcended, not, as Wallas proposed, from the momentum of rational effort that is released **after** the problem is solved.

Furthermore, Wallas allows his Incubation stage to remain in the shadows. For him, it is the "un" stage that is no longer conscious thinking but not yet a transformation into conscious insight. We have discussed that this stage, lying outside the reach of rational thinking, allows for unconscious parallel processing of information. Yet, our research reveals that this stage is more than the unconscious; it is meta-conscious and super-psychic as well, a realm fully available to Nihilitive Knowing. And finally, we propose that scientific insight is the elated thought born upon **exiting** this stage. Thus, speaking more correctly, the eureka! of science is actually an outsight, not an insight at all.

The Pattern of Knowing

We discussed earlier how Kurt Gödel's research skill of balancing insight and rational analysis is essential to the effective use of Nihilology as an interpretive tool. Yet now we will argue that we must go beyond even this sort of cognitive skill if we are to grasp the import of Nihilology as a guiding heuristic. We must also learn to be at ease in the study of three fields of research simultaneously. For this is one of the important new insights of Nihilology: in the synthetic pattern of science, religion, and philosophy we gain a deep and comprehensive understanding of Nothing, in clear differentiation from nothingness, and gain guidance in critiquing and enhancing the research in all three fields.

There is a story that finds its birth in Rodney's history of teaching at university. On the first day of a class in Christology, Rodney strode to the podium, velvet trimmed robe aflutter, and began to toss a purple rubber ball into the air, first up and down in the right hand, then up and down in the left, then from right hand to left. He stayed silent as students milled about finding seats and friends to sit with. He then added a second ball to the mix, a blue one, tossing the two balls up and down in the right hand, then up and down in the left, then from right hand to left. He stayed silent as students began to notice and whisper among themselves. Rodney then added a third ball to the mix, a gold one, tossing the balls up and between his hands in the classic pattern jugglers call the cascade. He stayed silent as students became enrapt and finally quiet. He then spoke:

> "The first ball is Jesus the living Christ. It is purple to show his kingly quality. Since it was normally in the presence of Jesus that disciples experienced the presence of the Holy Spirit, or perhaps in vision as was the case with Saul as he became Paul, I bring in the second ball. It is blue like the sky out of which the spirit seems to derive. And since Jesus said he was the way to the Father, I bring in a third ball. It is gold like the eternal precious metal.
>
> "To be a juggler, I must find a particular knack of visual perception. For if I focus on any single ball, the purple or blue or gold, perhaps because one is my favorite color... or if I rapidly shift my attention from one ball to the next, to see the three balls as individuated units, perhaps as an ante-digital attempt at multi-tasking, the juggle will collapse into a chaotic bouncing on the floor. The knack of perception I need as a juggler demands a flip, a somersault of attention to the **pattern** of balls, to the movement that ties the balls together.
>
> (And then his pedagogy hit its mark): "This is the sum of today's lecture. The Holy Trinity can only be appreciated in the pattern of the three persons: the purple Christ, the

> blue Holy Spirit, and the gold Father. As individuals they fall in a chaotic bouncing. In their cascade do they reveal their truth."

Just so, it is the **pattern** of evidence brought to light in the simultaneous study of multiple fields that many scientists today are coming to appreciate. They are the researchers who ply their scientific method at the juncture of brain and consciousness, science and theology, physics and metaphysics. They, by necessity, need an agility of thought that comprehends and values both realms of their research. A telling example of this breed of scientist is Andrew Newberg, MD, who began over twenty years ago to focus his research at the meeting point of brain and sacred experience. In this work, coined by him and his long-time colleague Eugene d'Aquili as Neurotheology, Newberg employed single photon emission computed tomography (SPECT) to document the neurophysiological mechanisms associated with religious experience with Catholic nuns at prayer and spiritual experience with Buddhist meditators.

The personal and intellectual character of scientists does affect the quality of their science. Before, it was described how when Rudy Rucker convened his first interview with Kurt Gödel, he was intimidated, expecting a strict European don. Yet, Gödel showed himself to be a joyful and "twinkly" character. John Horgan, famed as the author of *The End of Science,* was also pleasantly surprised at Newberg's person when he met him for a first interview. He was very young, with an eager-to-please attitude, and "too normal too to be delving into such esoteric affairs." As the interview progressed, Horgan became quite disarmed by Newberg's cordial earnestness and modesty about his groundbreaking research.

In the final sections of our book, where popularizer scientists in the field of cosmology are critiqued using Nihilology, we argue that humility is necessary in the search for scientific truth. We shall argue that hubris on the part of the popularizers and their imagined superhuman status in today's culture actually inhibit their scientific understanding.

With that, let's return to the actual research of Andrew Newberg. The SPECT technique he uses is a safe, non-invasive way to measure brain activity. It works because when a certain region of the human brain becomes active it needs more oxygen, and to meet this increased demand, blood flow increases to the active area. SPECT technology detects the changes in blood oxygen and displays them in the now famous, brightly-colored moving images on screen.

In his book *Why God Won't Go Away: Brain Science and the Biology of Belief*, Newberg discusses one particular experiment he performed on a subject named Robert who reports that, during peak periods of his meditation, he has a sense of timelessness and perceives the infinity of the universe. In those periods, he feels at one with everyone and everything in existence. In response to this report, Newberg states that, to the ear of a traditional scientist, phrases like "perceives infinity" or "one with everything" ring hollow, however "years of research have led... [me] to conclude that experiences like Roberts are real and can be measured and verified by solid science."

It must be made very clear that, by the term "real" Newberg only contends that Robert's experiences are neurologically so; they register consistently on the SPECT screen. And though Robert's experience is confirmed to be valid in his neurological system, it doesn't necessarily point to the reality of oneness or any other sacred realm.

Now Newberg shows his great talent of appreciating the **pattern** of findings in science and religion.

He reads research on near-death experiences and mystical states induced by epilepsy and hallucinations triggered by electrical stimulation of the brain. Beyond these scientific works, he also reads the writings by the mystics of world religions and mythologies, finding that "... important pieces of the puzzle came together and meaningful patterns emerged. Gradually, we shaped a hypothesis that suggests that spiritual experience, at its very root, is intimately interwoven with human biology."

Andrew Newberg is an exemplar of the skill of working in multiple fields of study simultaneously, and he has brought the new field of neurotheology into being and into respect among a wide reviewing audience of scientific colleagues. As stated before, working in multiple fields simultaneously is a necessary skill for Nihilology as well. In addition, Newberg shows deep interest in pursuing science to its limits and then reaching beyond science to discover what can be gleaned when rationality and empiricism comes to their inevitable impasse, which also epitomizes the research principles of Nihilology.

Just so, Nihilology appropriates from science as far as it has been able to go with respect to the language and insights into the question of Nothing. But then it goes farther, enriching the science with what has been grasped by theologians and philosophers alike. And, as we enter into a discussion of the theology and philosophy of Nothing, let me presage one essential perspective that will be revealed therein: regarding our explorations of Nothing, there must always remain something elusive, beyond rational comprehension; something that we should be satisfied and fulfilled at not being able to comprehend.

Gustavo came to this rather comforting conclusion some years ago while doing research for his history of science teaching. As he was looking into Anselm of Canterbury's infamous ontological argument for the existence of God and its implications for the empirical sciences he came across this statement in the *Proslogion* that stimulated an insight, "Therefore, O Lord, not only are You that than which a greater cannot be thought, but You are also something greater than can be thought."

After reading of this passage, Gustavo was surprised to hear himself reflecting, "It is truly fine to think **about** God, as this is the nothingness of God. Yet, this is merely a thinking **around** the Nothing of God, which is his actual nature." He was at peace with this reflection because it freed him to believe, to have faith in God, knowing that his thoughts can only take him so far.

One final note in this discussion. it is in certain areas of the exploration of Nothing effective to think of religion as an overarching approach to the domain of reality that lies beyond the bounds of rational thought. Religion offers human beings a host of ways (*via*) to leap into and stay awake in this meta-conceptual domain: Apophatic Theology, mindfulness of the constituents of selfhood, confronting the conundrums (*koan*) of the master teacher, contemplation of God's creativity (*sefirot*), and faith. They all aid toward the goal of Nihilitive Knowing. In later sections of the book on the epistemology of Nothing, we will expand on these ways. For now, it is only important to note that religion, so defined, can be a guiding rule of thumb, a micro-heuristic in the wider tool kit of Nihilology.

(3)
The Nothing of Religions

In the beginning, it should be said that we do not intend to burden readers with a wide-ranging history of world religions. Many other scholars with glossier credentials and deeper insight than ours have done extensive writing to that end. Yet, remember our disclosure about how we began to look closely at the history of science and what a shock it was to see Nothing everywhere and through time? During that same research, we hoped to let there be light on Nothing in the history of religions as well.

We did not have to look far. The second verse of Genesis, the very first chapter in the Hebrew Bible, employs this wonderful mythic language to give us access to the condition before creation of the universe, "The earth was a formless void and darkness covered the face of the deep…" (Gen. 1:2a, NSRV).

Interesting how these ancient writers were so playful in presenting us with such weighty matters as how the world came to be. "Void and without form" translates from the Hebrew *tohu va-vohu. Tohu* carries the weight of denotation here, meaning a void of unstructured chaos.

Vohu is called on, in the main, as a rhyming word. The Torah is traditionally a chanted text, and the sound of scripture carries as much import as the denotative meaning. Yet, why would those ancients open their most sacred text with a phrase akin to "chaos and anarchaos" or "nothing and naughting" or "void and emptoid"? It seems that the fear of the void was so intense and widespread in the ancient Mideast that a somewhat silly irony was intoned to sooth the minds and hearts of the cantor's rapt listeners.

So, as it turned out with more research, religions throughout the world have been just as intrigued—and terrified—as science in the pursuit of Nothing. Let us take a close look at four themes and apply their findings to the building of our heuristic Nihilology. Along the way, we will present specific parallels between these religious/theological themes and the theories of science which may aid readers in making sense of the sometimes-esoteric teachings of religions. (For some readers, the theories of science are esoteric unto themselves!) Finally, let's start in the East for each theme and then travel back home to the West. Again, this should help readers understand the themes of religions and help them apply that understanding to the use of Nihilology in their own studies of Nothing.

Nothing is Ever-present

Buddhism arose as a religion in the sixth century BCE in the extreme north of India, along the border of today's Nepal. It arose in a prince of the Shakya clan who renounced his family and royal power to pursue a life in the jungle as a spiritual hermit. There was a profound strain of antipathy to worldly life during this period. As an example of this, the mythology has it that the son of the Shakya prince bore the name Rahula, which means "fetter" or "snare."

This Shakya prince became the Buddha when he sat, unmoving, beneath a tree for six days that culminated in the explosive understanding that the soul (*atma*) does not indeed exist as the root existential fact of humanity, as had been taught to him by his forest mentors and by the Hindu theologians of his day. Buddha's sermons on *anatma* bloomed into the scriptures of the Theravada Buddhist tradition that held sway for several hundred years after the Buddha's death.

Then something momentous happened in the realm of Buddhist thought. In his book *The Heart Sutra: A Comprehensive Guide to the Classic of Mahayana Buddhism* Kazuaki Tanahashi describes how the arising of the monumental body of texts called *Prajnaparamita* about four hundred years after the Buddha's time served as the root catalyst for the rise of Mahayana Buddhism. "These scriptures served as a huge step whereby the early notion of *anatma* (no-self) developed into *shunyata*—an interrelated non-existence of individual entities, as well as of all phenomena."

Buddha preached that the human being is without a real self. Analyze the human. One will find feelings, vital organs, a name, memories, and desires, but no selfhood. Then the later Mahayana Buddhists expanded this notion of no-self to make what is perhaps the most radical statement in theological history: all manifest life is without inherent, independent existence. Analyze a chariot (Warriors drove them in north India during this period), you will find wheels, a yoke, a name, an axle, but no "chariot." The human **and** the chariot are *shunya*.

The texts in which *shunya* functioned as a root notion are called *Prajnaparamita*, literally "having crossed over to the other shore [that is] wisdom." Crossing over to the other shore is an old Buddhist image of riding in a boat across the river of ignorant life to find nirvana on the other side. Over time, the phrase came to mean "perfection." One becomes perfect in seeing the *shunya* quality of all manifestation.

It should be noted that, in the *Prajnaparamita* literature, *shunya* [nothing] was used interchangeably with *shunyata* [nothingness]. The simple noun or the abstract form were called upon as linguistics or syntax necessitated. These Buddhist writers had no interest in distinguishing between Nothing and nothingness. However, in our development of Nihilology as a heuristic tool, this differentiation is essential. So, in the following, the words *shunya* and *shunyata* will both be translated as "Nothing." In doing so, we maintain an honest loyalty to the Buddhist intent while also maintaining a consistent approach to vocabulary throughout this book.

The most succinct—and famous—elucidation of *shunya* in the *Prajnaparamita* literature is found in the Heart Sutra. ('Heart' translates from *hridayam* which also carries connotations of "the heart of the matter" and "pith.")

Form is nothing... nothing is form
Nothing is not separate from form... form is not separate from nothing
Whatever is form is nothing... whatever is nothing is form

"Form" is one way that Buddhists denote the manifest universe, and their unique insight is that this manifestation is not real *qua* manifestation. The manifest universe is real only in as much as Nothing is present within it. Once again, Kazuaki Tanahashi helps us by explaining that there are two main sorts of world views proposed throughout religions of the world: pluralism and singularism. Pluralism analyzes in a practical and intellectual manner the phenomena of life according to their difference from one another. Whereas Singularism is a non-dualism that incorporates the pluralistic worldview by going larger to a vision of reality as a unified whole that contains all relativistic phenomena. Singularism is the major foundation for Buddhism.

Further, in the fifth century of the Common Era, about eight hundred years after the death of the Buddha, a particular school within the Mahayana tradition called *Dhyana* (meditation) migrated across the sea to the southern shores of China. The teachings spread north, and with Chinese ears hearing the word *Dhyana* in their own particular way, the school became known as *Chan.* By the eighth century, the school had made its way to what is now known as the Korean Peninsula and across the sea to Japan. The Japanese heard the word *Chan* in their own particular way, and their school became known as Zen.

For our exploration of the Nothing of religions as it aids us in building the heuristic of Nihilology, this Zen form of Mahayana is especially important. Zen posited that Nothing is directly perceivable, with enhanced attention, in our human encounter with everyday objects. In a section below on epistemology and Nothing, we will expand on the seeming impossibility of knowing Nothing. For now, let us focus on the ontology of Nothing.

Mumonkan is one of the two classic collections of *koan*, those existential puzzles that present students with intellectually unsolvable paradoxes. In *koan* number fortythree, we find a pithy traditional story of the tenth century Zen master Shuzan. Sitting before a group of his disciples, he holds up his bamboo stick (*shippei:* the symbol of his teacher status) and challenges his students. "If you call this a bamboo stick, you oppose its reality. If you do not call it a bamboo stick, you ignore the fact. Now what do you wish to call it?"

The Zen form of Mahayana is steadfastly dedicated to an indigenous strain of minimalism and simplicity expressed in many ways throughout Japanese cultural history. Shuzan, a master of the straight-forward, stimulates his students into a condition of heightened attention simply by holding up his symbol of authority. Then, if a student calls the bamboo stick a something, they falsely assert its independence from all other things around it and from its ground in Nothing. Yet, if a student does not call it a bamboo stick, they deny any existence to it at all.

In terms of human experience, the Zen school teaches that any manifest object (Shuzan's stick just happened to be in his hand at a teachable moment) must be perceived "surrounded" by Nothing. One might say that things float in Nothing as their context and this context lends reality to the thing.

Manifestation is not in any way separate from Nothing. Both manifestation and Nothing are present at all times, even during manifestation. Thus, manifestation, in and of itself, is not genuine. Manifestation, unseparated from Nothing—or should we say the unity of manifestation and Nothing—is genuine, the only genuine. Thus, we harken back to the verse from the Heart Sutra. "Form is none other than nothing, nothing is none other than form."

Now, it was stated before that the *Dhyana* form of Buddhism came by sea to China. When these Buddhists arrived on those southern shores, they met with the indigenous scholars of Taoism, and the *Chan* school was born of these conversations. For the purpose of building Nihilology, it is important to emphasize that these Taoists had their own unique and insightful notion of Nothing.

As an interesting cultural aside, in China the stories of a Buddhist sage, a Taoist sage, and a Confucian sage talking of deep matters around a tea pot function as an eastern equivalent of the western, "A priest, a rabbi, and a minister walk in to a bar."

The mythology of Taoism tells that Lao Tzu gave forth its earliest teachings in the sixth century BCE, making him a contemporary of Buddha. ('Lao Tzu' translates as "old teacher." This is a title of honor, not a personal name. Most likely, the old teacher is a mythic character.) These teachings have survived through time in the tradition's urtext: *Tao Te Ching*. Stanza 11 reads:

> Thirty spokes are joined together in a wheel,
> but it is the center hole
> that allows the wheel to function
>
> We mold clay into a pot
> but it is the emptiness [*wu*] inside
> that makes the vessel useful
>
> We fashion wood for a house
> but it is the emptiness [*wu*] inside
> that makes it livable.
>
> We work with the substantial,
> but the emptiness [*wu*] is what we use

As with most ancient Chinese texts, the *Tao Te Ching* is always accompanied by a long history of commentaries that help us moderns gain access to the subtle meanings of the original. For instance, Li-Jing comments, "It is because the hub is empty that the spokes converge on it. Likewise, it is because the minds of sages are empty that people turn to them for help." Ch'eng Hsuan-Ying follows by saying, "A cart, a pot and a house can hold things because they are empty. How much more those who empty their minds."

The central terms being juxtaposed here in stanza 11 and its commentaries are *wu* and *yu*. *Wu* carries the connotations of "empty," "the not," and "nothing." *Yu* connotes "something," "the made," and "created." In our lexicon of Nihilology, these correspond to Nothing and "creation." And here is the first essential point to be learned from the Taoists, one that echoes and extends the Buddhist notion of form and Nothing discussed before: Nothing and creation are eternally present. Creation is not the end of nothing. Nothing is not lost at the creation of the universe.

The second point is that Nothing gives creation an isometric balance that allows created things to function. The first verse of stanza 11 of the *Tao Te Jing* tells us that a wheel can roll a cart down the path due to its empty hub. It is, of course, possible to manufacture a solid wheel, yet the ideal wheel would not be solid because it would be too massive. Rather, it would be a thin frame enveloping the hub via a spider web of spokes, so as to make the wheel as light as possible.

This principle has shown itself in the history of western scientific ideas; the function of form, i.e., created things, is more and more revealed by the shedding of excess matter. Let us call it the **rarefying** principle and see that it has been particularly active as scientists have attempted to answer the question, "What is the manifest world made of?"

In an earlier section of our book, we discussed how Sir Joseph John Thomson, at the turn of the century, proposed a model of the atom as negatively charged corpuscles floating about in a sort of ooze carrying a positive charge. Then soon after that, around 1910, Ernest Rutherford discovered that Thomson's ooze did not actually exist. Rather, those corpuscles, renamed electrons, were negatively charged particles "flying" about a massive positively-charged center and that, true to the rarefying principle, most of the atom was indeed empty space.

This image of the atom, with the addition of neutrons in the nucleus, held sway in western scientific literature until the 1960s when it was discovered that protons and neutrons are not elementary particles at all. They are themselves composed of quarks.

Two of the six quarks are termed "charms" and "strange." Thus, we see the language of reality's physical make up get more and more rarefied as its constituents become more rare as well. Another example of this: quarks come in "flavors," not more substantial sorting categories like types or species.

These quarks are surrounded by space within the larger particles of neutrons and protons. Thus again as a created thing becomes more and more understood, it turns out to be lighter and lighter; it sheds matter.

Then, in the 1970s, an even more profound rarefaction occurred. Up to this time, all particles in the atom were envisioned as point-like in nature. Scientists began to posit that particles are not points at all; rather they are vibrating strings. (These strings have only one dimension and so defy our notion of matter taking up three dimensions.) What scientists had taken to be distinct particles were now simply different rhythms of oscillation. Matter had been reduced to rhythm. Very rare indeed!

Now, let us return to the *Tao Te Ching*. The third verse of stanza 11 gives us another image of Nothing making creation work. It says that people can actually inhabit a house, not because of the walls and roof, but because of the empty space that the walls and roof enclose.

Again, there is a parallel to this Taoist concept in the annuls of western thought, this time in architectural literature as brought forth by the wide-ranging thinker Frank Lloyd Wright. (In the following, just think of the created universe as a house.) In his famous book from the 1950s *The Natural House* he stated, "The reality of a home is to be found in the spaces enclosed by the roof and walls, not by the roof and walls themselves." This recognition of the essential presence of space was a radical insight for Wright, and it developed into a radical break from the architectural theory that he had been taught. He titled a 1952 speech to young architects "The Destruction of the Box", boldly announcing this break. In that speech he declared what many scholars read as a metaphysical tenet, "The space within the building is the reality of that building!"

And Wright went beyond this basic statement about materials and the spaciousness enclosed. As explained by Gail Satler in her book *Frank Lloyd Wright's Living Space* "[Wright's] organic architecture reveals space as a lived in and a living entity, one that achieves its full meaning only when it becomes inscribed with the actual practices of those who inhabit it." This notion that the people within a space bring that space to life and that the space and the people find meaning in this relationship is a statement that could easily have been uttered by a Taoist teacher.

There is also a third point that the Taoists make about *wu*, i.e., when the mind is empty of concepts, *wu* or Nothing can be known. And, as a fourth point, the term *wu* is used in a related but clearly differentiated way in the most famous phrase in all of Taoism: *wu wei* [action-inaction]. We will contend with both these usages below in our discussion of the epistemology of Nothing.

So, to conclude this theme, we have gleaned two central notions regarding Nothing from the teachings of Buddhism and Taoism. First,

Nothing is always present within the manifest universe. Second, the things of the universe are made functional by the always-present Nothing that lies at its center. As we shall see later, this view differs from that of the popularizer cosmologists who see Nothing as an absence, at a "distance" from, and not permeating the universe after the Big Bang of creation.

Nothing and nothingness

It was stated before that in building the heuristic of Nihilology it is essential to differentiate between Nothing and nothingness as they apply to the universal nature of existence, to human beings and the material creation. As we turn our attention to gleaning from the religions of the world, we realize that the most fruit for our purpose can be picked from doctrines—eastern and western—that give voice to the *via negativa,* the way of negation. The urdoctrine of the negative way is that it is not possible to make a true statement that encompasses the nature of Nothing. As we have seen, Nothing is given the name *shunya* or *wu* in the East. And, as we will propose, Nothing is given the name God, *Ayin*, or Allah in the West.

A paradox might come immediately to mind for readers: if a true statement cannot be made about the nature of Nothing, how is it that teachers of the *via negativa* throughout the world and time have been speaking and writing volumes on the subject? Chris Boesel and Catherine Keller express the paradox well in their book *Apophatic Bodies: Negative Theology, Incarnation, and Relationality.* They state that the purveyors of the *via negativa* do surely speak, however, "The more they speak, the more they unspeak; and yet because of the infinity of which they speak, it would seem they can never stop speaking..." Those of the negative way seem to become speechless for a moment before the very same mystery that inspires speech in the next moment!

As can be seen in these quotes, any discussion of the *via negativa* invites a sort of self-contradictory language. Yet, it is important not to allow our analysis to succumb to a facile word shuffle, for teachers of the negative way speak and write is an important one: the *via negativa* is a **way.** Speaking and writing, as well as listening and reading, about Nothing is held to be an effective means of transcending the very words uttered. (We will present our understanding of this transcendence below in the section on the epistemology of Nothing.)

The stellar voice of *via negativa* from the East is the third century Buddhist scholar Nagarjuna. He was so central to the birth of the *Mahayana* (Great Vehicle) school that adherents in Tibet and China often referred to him as the Second Buddha.

By the time of Nagarjuna, Buddhism had become an intricate tradition of philosophical positions, debate structures, and centers of learning. Nagarjuna himself earned his fame as a debater in what is now Bihar, India at Nalanda University, the greatest center of learning in the history of Buddhism.

Now, Nagarjuna clung staunchly to the *via negativa.* He would only agree to a philosophical debate in the role of a *vaitandika* (destroyer), committed to destroying all opponents' theological and epistemological positions without ever proposing a positive assertion of his own. The particular lineage that Nagarjuna gave birth to was called *Prasangika (*Adherents of the Unwanted Consequence). In the Western philosophical lexicon, this "unwanted consequence" is known as *reductio ad absurdum.* Nagarjuna argued within the roles and rules his opponents accepted and reduced their positions to absurdity.

Nagarjuna, the debater, took seriously two central Buddhist "destroyer" tenets: We as human beings can never establish a stable essence in any created thing of the world. Things unto themselves are *shunya* (Nothing). Also, created things are in a condition of constant change and only exist for our experience having heeded other things. This is the famous tenet known in the texts as *pratityasamutpada* (dependent origination). So, Nagarjuna, bound by his opponents' rules, did not actually strive to win the contest but strove to expose the contest itself, which all along had been a sort of practical, efficient religious exercise without defensible, real-life referents.

This "efficient religious exercise" is the **way** aspect of the *via negativa.* Mark Siderits and Shoryu Katsura, in their book *Nagarjuna's Middle Way,* tell us the aim of the destroyer is not to shock the opponent but rather to employ eight allowed debate negations to foster nirvana by bringing an end to hypostatizing (*prapanca*). "By hypostatization is meant the process of reification or 'thing-ifying,' taking what is actually just a useful form of speech to refer to some real entity."

Nagarjuna's debate undermines the opponents' experience of created things as separate from Nothing. His debate undermines the opponents' capacity to hold to any ultimate statements about created things. This aids in bringing about nirvana, the ultimate state for all Buddhists, the state in which the self and created things are seen to be one with Nothing.

Now, Nagarjuna was not only a great debater but was also the author of the most famous text of his school: *Mulamadhyamakakarika* (Root Verses of the Middle Way). *Karika* are pithy poetic statements intended as memory devices for students to quote back to their mentors for elucidation. As such, they contain the distillations of the school's teachings. As before, we will translate *shunya* and *shunyata* as "nothing."

1. I prostrate to the Perfect Buddha
The best of teachers, who taught that
Whatever is dependently arisen is
Unceasing, unborn

2. Unannihilated, not permanent
Not coming, not going,
Without distinction, without identity,
And free from conceptual construction

14. Whatever is dependently co-arisen
That is explained to be nothing
That, being a dependent designation,
Is itself the middle way

54. What language expresses is nonexistent
The sphere of thought is nonexistent
Un-arisen and un-ceased, like nirvana
Is the nature of things

30. The victorious ones have said
That nothing is the relinquishing of all views
For whomever nothing is a view
That one will accomplish *nil*

52. Action and misery having ceased, there is nirvana
Action and misery come from conceptual thought
This comes from mental fabrication
Fabrication ceases through nothing

These stanzas are sated with language of negation, just as were Nagarjuna's words in his debates. Nothing is "free from conceptualization." Whatever is expressed in language is "non-existent." Mental fabrication "ceases through nothing." In sum: Nothing is not within the reach of conceptualization. This is true *via negativa*, yet what is the nature of that which **can** be spoken of, written about, and thought upon?

Let us turn to one of the most important spokespeople of the modern era of Buddhism to gain some clarification in this regard. Chögyam Trungpa Rinpoche was a Tibetan lama who led in bringing Buddhism to the west in the 1960s onward. He possessed an extremely subtle sensitivity to the sound and idiomatic usage of the English language. (His school of Buddhism in the US created a T-shirt that went viral in the 1980s. The logo read "Meditation: It's not what you think.") His approach to teaching Nothing (*shunyata*) came to be because of the mindset of North American students who were hyper-intellectualizers and needed the reminder that they could not cogitate themselves to Nothing.

In his book *Glimpses of Shunyata,* Trungpa opens a seminar by warning that, though the purpose of the gathering is to understand *shunyata* or Nothing, this "understand" is suspect. "We expect some concrete answer, something definite, something solid, but solidity itself depends on frivolousness, so to speak. The idea of *shunyata* depends on what is not *shunyata*——which is based on ego's manifestation."

The teaching of Buddhism is very clear here. The **idea** about Nothing is not Nothing itself. Ideas form around Nothing. Ideas are not about Nothing at all. K.C. Cole, as always, is helpful in her straightforward advice for people who have taken on a meditative discipline toward acquiring a sense of Nothing. "But to think about it—to analyze its characteristics—requires that kind of structured thought that by definition destroys the very nothing you seek."

A parallel sort of paradox from the history of science may aid us in our understanding. It is well known that elementary particles have a dual nature, sometimes wave and sometimes particle. For example, when electrons are emitted one at a time through a double slit screen, they act as particles until they strike the screen and act as waves when they exit on the other side. However, and herein lies the paradox, if an observing mechanism is set up to see which slit the electrons actually pass through, the exiting electrons no longer act as waves. The scientific literature has termed this phenomenon the "collapse of the wave function." And while a conversation is still going on among scientists, it has been generally accepted that the very act of watching the behavior of the electrons alters the nature of the electron itself.

According to the view of the *via negativa,* as we have witnessed thus far in the teachings of Buddhism, it is very much the same regarding the act of thinking about Nothing. Any conceptualization about the nature of Nothing alters its nature. It is no longer Nothing about which we are thinking. Thinking is, indeed, **about** nothingness, only a thinking **around** Nothing. For the purpose of the heuristic of Nihilology, let us encode this phenomenon thus: all concepts about the nature of Nothing are actually about nothingness.

Employing the abstract form of the noun is apropos in our investigation as it carries much denotative and connotative weight. "Nothingness," linguistically, refers to the entire class of nothings, a summary of all experience we ever had with entities that can be empty, for example, water tanks, begging bowls, or the night sky. The term allows us to speak in generalizations yet without a specific referent. Nothingness, in the language of ontology, is a state that cannot be experienced with the five senses of taste, touch, hearing, smelling, or seeing. It is only a creation of thinking.

So now, with this basic differentiation between Nothing and nothingness established from the Eastern Buddhist traditions, let us turn to the expressions of the *via negativa* in the West to fill out this difference and gain a deeper understanding of why this differentiation is important in Nihilology.

From the early history of both the Orthodox as well as the Roman Church, the negative way has been given voice in what their thinkers over time coined Apophatic Theology. The term *apophasis* comes from the Greek of the Hellenistic Period and means "to deny."

The great schism beginning in the eleventh century that ended up dividing these two strains of Christianity, both of which claim to be "Catholic," was much more about crusading knights sacking Constantinople and the use of unleavened bread for the Eucharist in Roman liturgy than it was about theological differences.)

The essential assumption of Apophatic Theology is that God's essence is ineffable, and therefore we human beings must accept that language is inadequate to truly capture God. We should note that there has been, just as ancient and expressed in both the Orthodox and Roman Catholic Church, another voice in Christian history: Cataphatic Theology. Its adherents have sought to understand God in positive terms. They emphasize that God has been fully revealed in the person of Christ and in the Bible, which is believed to be inspired. Therefore, the preaching of Christ and the words of the Bible do, according to Cataphatics, indeed fully capture the essence of God. Of these two voices, it is only that of *apophasis* that has cared to, or had the subtlety of mind to, differentiate between Nothing and nothingness. Thus, Apophasis is akin to the Nihilogical voice we are discovering, and thus, we will look further into its words of denial and leave affirmation for others to explore.

Now, it was in the person of John of Damascus, the renowned eighth-century thinker from Syria, that we hear the clearest and most articulate expression of Apophatic Theology out of the Orthodox church. In his central work *An Exact Exposition of the Orthodox Faith* he teaches that no amount of theologizing reveals God's essence [*essentia*], "rather, what He is not... [Affirmations reveal] not the nature [*naturae*], but the things around the nature."

Just as in our discussion before where we saw that Buddhists hold ideas about Nothing (*shunyata*) that are not the actuality of Nothing, so we see here in Orthodox Christianity the exact same teaching regarding God. In this we begin to get a sense of the confluence of the theological notions of God and Nothing.

Before going farther, it is quite revealing to note that, while holding a somewhat fundamentalist grip on the *via negativa* regarding words, John of Damascus simultaneously spearheaded an infamous theological battle with the leaders of the Orthodox Church to fight for the use of images. In 726 CE, the Emperor Leo III banned all images from the Orthodox liturgy. Church officials began a smashing campaign, an iconoclasm. John fought back by saying that when one contemplates the incarnation of Christ it is acceptable to clothe him in human form. "[Since God] empties Himself and takes the form of a servant in substance and in stature and is found in a body of flesh, then you may draw His image and show it to anyone willing to gaze upon it."

For our purpose of building Nihilology, it is the following statement from John of Damascus's central work in this fight, *Defense Against Those Who Oppose Holy Images,* that is most intriguing. "The invisible things of God have been made visible through images since the creation of the world."

While emphasizing the *apophasis*, the denial, of words to capture the nature of God he says, "These do not show what He is, but, rather, what He is not." And he upholds the positive efficacy for images to reveal God. "The invisible things of God have been made visible..." This differentiation by John of Damascus between words and images is supportive of our discussion above about images serving as trans-heuristics in bringing about scientific insight. Again, we begin to see the confluence of central notions in Nihilology, the experience of scientific insight and the experience of God. Below, in the section titled Nihilitive Knowing, we will make a good case for how this is so.

One of the great voices of the Roman Catholic Church, John Scotus Eriugena, helps us make subtler our understanding of Apophatic Theology. Eriugena, whose name means "born in Ireland," became lauded throughout the Church for his Greek Scholarship. It was a rarity in his day for an outlier from western Europe to know Greek, and he was taunted about his birthplace by the ninth century elite scholars of Rome. Nonetheless, he grew to prominence among theologians through his commentary and translation into Latin of the works of the Greek scholar known as Pseudo-Dionysius. In his *Patrologia* he teaches that we do not know what God is and God himself does not know what He is because He is not any thing. "We may learn from creatures *that* God is but not *what* God is.

In his five-volume master work *Periphyseon,* Eriugena divided the entire universe into various natures (*naturaes*). Anything we can sense or think about was said to **be,** whereas that which is beyond experience via sense or thinking was held **not** to be because of the excellence of its nature (*per excellentiam suae naturae*). In Eriugena's world, because God's nature is so excellent (*nihil per excellentiam*), He is **not**.

These are indeed the bald denials of the *via negativa.* Even God himself does not know what his own nature is. This negation is not intended to limit God, rather to negate the power of theological understanding. In the language of Nihilology, God is Nothing. Theology can convey to us that God exists, that is God as nothingness, but it cannot tell us His nature.

Now, Eriugena goes much farther than other theologians in making the subtle point that the form of theological speech does not have to be negative in order for the *via negativa* to find its impact. For example, he calls out his listeners with the enticing statement that God is *superdeus deitas,* the divinity above the divine. Or in Comic-Con language: Super-God! Frederick Copleston, in his *A History of Philosophy* explains that Eriugena affirms God as "super wisdom," and there is no linguistic negation here, yet the human mind experiences a negation of content. "The *via negativa* is thus fundamental, and as we do not pretend to define *what* the 'super' is in itself, the ineffability and incomprehensibility of the Godhead is unimpaired."

Perhaps, a parallel from the history of science would help us get a sense of Eriugena's "super." Niels Bohr received a Nobel Prize in 1922 for his work in creating the beautiful and elegant model of the atom that many of us grew up with in science classes. It depicted a "solar system" in which a nucleus at the center is encircled by electrons flying in discrete orbits with discrete energy levels. And, just like in our macro-solar system of planets, the void in between the electron orbits is much vaster than the inhabited space of the electrons themselves. It is this void that actually gives form to the atom. Just so, we may say, for Eriugena, God is such a super void.

So, let us put a cap on this particular phase of our discussion with a prayer from the thirteenth-century theologian known as Meister

Eckhart, who was profoundly in debt to Eriugena's teaching. Though he was called before the Inquisition for his mystical preaching about his oneness with God, Eckhart has been remembered as a most humble servant, as seen in this, a masterful statement from one of his sermons about Christian Apophatic Theology, "I pray God to rid me of God."

In this epitome of pith, Eckhart is using the term 'God' in both senses, as we defined above: "God (Nothing), please rid me of God (nothingness). Allow us the nerve to restate his prayer: "Oh infinite God who is real and beyond all my conception, please rid me of all limited ideas about you."

Here Eckhart is differentiating between Nothing and nothingness. If God is beyond concept of any sort, all talk about God, even words of negation, is talk of God as nothingness. So, this ancient tradition of Apophatic Theology both says and unsays talk about God.

Now, as was discussed before, the *via negativa* is a **way**, a means to fulfillment, in this case a way to God for Christian believers. In culminating this phase of our discussion on Nothing and nothingness, here are two more perspectives on how this way works. Chris Boesel, editor of *Apophatic Bodies: Negative Theology, Incarnation, and Relationality* says, "For, without our finite images, images of the illimitable drawn from our bodily limits, would 'God' exist *for us*?" This question is utterly essential to the religious lives of most Christian adherents. The every-Sunday members of the flock have no desire to let go of the nothingness of God. The words of theologians and clergy remain most integral to their understanding of Him. And, in order for them to be able to fulfill their religious duty, "God" provides the only tool to work with.

In contrast, Boesel also states, "The apophatic mystics... do surely speak.... The more they speak, the more they unspeak; and yet because of the infinity of which they speak, it would seem they can never stop speaking." These lines tell us of yet another aspect of the *via negativa* as a way. Apophatic Theology is often associated with mystic strains of Christianity. To understand this endless talk of God, we need to ask again, as we did before, this question: why do those of religious experience bother to speak?

We know that words have denotative meaning and carry connotative meaning. Yet, science now has shown that words also have effect completely outside of meaning. Neurotheology has proven that words, both heard and thought internally, can alter the makeup of our brains and affect how we perceive reality.

In their book *Words Can Change Your Brain*, Andrew Newberg and Mark Robert Waldman tell us that spoken language impacts our mind in ways that seem impossible to the lay reader at first, yet with deeper study, the research becomes quite convincing. For instance, they show that a single spoken word has the power to influence how genes express to regulate physical stress and emotional stress.

Regarding words in the form of thoughts, they found if we reflect in a sustained way on the nature of God, the brain will be altered in those centers that regulate mood, create conscious experience of selfhood, and even shape the way our senses interface with the outer physical world. For example, conscious thoughts held over time will restructure the thalamus, the information relay station of the mid-brain that guides consciousness and sleep, effecting an alteration in the way reality itself is perceived.

Newberg and Waldman crown this line of thought in a capstone statement from another of their partnership books *How God Changes Your Brain*. They lead up by noting aspects of their research: if one contemplates the notion of God for an extended period of time, neural functions in the brain alter. Alternate circuits activate; common circuits go quiet. Dendrites, or nerve branches, begin to grow. Synapses connect up, and the brain becomes more capable of new realms of experience. Then the capper "God becomes neurologically real."

Let us go one step farther in this neurology of words. Newberg has also found that the brain is uniquely altered in the midst of a concentrated conversation. When two people who have compassion for one another engage in extended talk, each begins to have the ability to predict what the other is going to say before it is actually voiced in the form of words. Both people's brains begin to align in a manner that neurologists have termed "neural resonance."

Steven Lehar, a neuropsychologist with leading edge insights into brain function says, "[What we normally refer to as] 'capacity for empathy' may be shorthand for 'ability to accurately simulate or mirror other peoples' neuroacoustic properties." This brain mirroring has also been observed as an active function—skill, one might say—in the field of counseling psychology. As Elsie Jones-Smith tells us in her book *Theories of Counseling and Psychotherapy: An Integrative Approach*, neural resonance occurs when a therapist listens to and observes a client as deeply and fully as possible. Brain research has revealed that the more deeply a therapist listens, the more their brain will mirror the activity of the client's brain.

From this discussion we find greater understanding of how the *via negativa* functions as a way. For the building of Nihilology, we can also begin to form a principle for why Apophatic theologians speak. They only speak of that which is experientially present at the moment of speaking. By this we mean that the speaker must be experiencing what they are speaking about as a prerequisite to their words having the effect of bringing the listener into that same experience. When a theologian resides in the experience of God (Nothing) and speaks to listeners about "God" (nothingness), those words point beyond themselves and function to bring the listener into the same experience. We will clarify and expand what is meant by "experience" in the section on Nihilitive Knowing. Yet, for now, we have a beginning in the findings of science that have us understand "same experience" to mean the mirroring brought about by the hearing and reflection upon words.

In the world of Zen Buddhism, there is a warning to help listeners understand, and make effective, this principle of speech, "Don't confuse the finger pointing at the moon for the moon!" The apophatic version says, "Don't settle for words about 'God' when God is present."

As a conclusion to this theme on Nothing and nothingness, let us turn to the Apophatic Theology of Islam and Judaism where we will glean more points to aid us in building our heuristic Nihilology. As we noted before, the *via negativa* is often most clearly expressed in the mystical forms of religious traditions.

The leading voice of Apophatic Theology in the history of Islam was Al-Ghazali, the Persian mystic who lived in the last part of the eleventh century. Much of his professional life was spent in Baghdad as a prestigious academic under appointment by the Vizier. The title of *Mujaddid*, Restorer of the Faith, was bestowed upon him. Over time, he gained the unofficial title of *Hujjat al-Islam*, the Proof of Islam, to which he is often referred even today.

Al Ghazali exhibited an eloquent voice in the service of *apophasis.* He stated that Allah is invisible and indivisible, unconfined by space or time, outside quantity or quality, and untouched by concepts of shape, color, or size.

Al-Ghazali, especially in his writing *The Alchemy of Happiness*, taught that these terms of *apophasis* apply to the nature of the human spirit as well as to Allah. "Not only are man's attributes a reflection of God's attributes, but the mode of existence of man's spirit affords some insight into God's mode of existence." The subtle intent of this "reflection" will be an essential part of the discussion in a later section of our book concerning Islam and, what we term, ways of staying awake.

Now, according to Islam, the greatest of the 99 beautiful names of God is Allah. There are at least two root meanings of this beautiful name. Allah is an Arabic compound noun consisting of *al* and *lah*. *Al* is a definite article translated as "the." *Lah* can be translated as "god." Thus *allah* can mean "the god." *Lah* also translates as "no" (as opposed to *naam* or "yes"). So, *allah* can also mean "the no" or "the nothing." Applying this bivalence to one of the most famous creeds of Islam, we learn much about its connotative power: *la illa-ha illa allah.* At one and the same time, it means, "There is no god except the God (Allah)" as well as "There is no god except the Nothing (Allah)."

With this in mind, we can better appreciate Al-Ghazali's example of the urdoctrine of the negative way. It is not possible to make a true statement that encompasses the nature of Nothing, in this case the Nothing of Allah. "God is great [*allahu akbar*] does not mean that God [Allah] is greater than creation... It rather means that God's greatness immeasurably transcends our cognitive faculties, and that we can only form a very dim and imperfect idea of it."

This is pure *via negativa*, but Al Ghazali goes much farther and gives us yet another view of the negative way as a way. He taught that, while living on earth, Muslims should learn about Allah via *apophasis,* calling this the "delight of knowledge." And this learning is essential to attaining a direct encounter with Allah in the afterlife, calling this the "delight of vision."

Al Ghazali had a gloomy attitude toward embodiment, calling it an imprisonment in clay that can encase humans in sensual experience which puts a veil between humans and Allah, concealing the vision of Him. However, while embodied, all humans can gain a sort of foreknowledge about Allah which may turn into a vision of Allah after physical death, a vision that is quite impossible without this foreknowledge.

An analogy from the world of computers may help us understand what is being taught here. According to Al Ghazali, the human being forgets everything at the moment of death, yet how the person prepares their mind before death is key for the vision of Allah when they re-awaken in the afterlife. This is akin to a computer that needs to perform a system update before shutting down. Only if the computer performs this system update will it be able to reboot and function in the upgraded capacity. Just so, according to Al Ghazali, a human will not acquire the "upgraded capacity" that leads to the delight of vision when reawakened in the afterlife without the foreknowledge of Apophatic Theology that, at minimum, instills the belief in the upgrade of life after death.

Al Ghazali, going even farther, adds one more necessary requirement to attaining an afterlife vision of Allah. In the tenth chapter ("The Remembrance of Death and the Afterlife") of his book *The Revival of Religious Sciences*, he quotes this conversation between Aisha, one of Muhammad's wives, and the Prophet himself. "Said Aisha (May God be pleased with her) 'O Emissary of God! Shall anyone be resurrected alongside the martyrs?' 'Yes,' he replied. 'He who recalls death twenty times in one day and night.'"

This injunction is *via negativa* at its sharpest. One is to bring to mind twenty times each day the very negation of the life one is leading. By negating the capacity of the mind to conceive of the true nature of Allah in conjunction with negating the capacity of the mind to indulge in the sensualities of living, one is prepared for the vision of Allah on the other side of death.

Al Ghazali also gives forth beautiful poesy regarding this vision in saying that one who turns their back on things of the earth and shows affection for God will welcome death when it arrives and welcome the union with Him whom he loves. In his case the Prophet's sayings will be verified. "Death is a bridge which unites friend to friend."

Before we move on to the final discussion in this theme, let us foreshadow our critique of popularizer scientists. Al Ghazali warned religious seekers who wallow in their experiences, pretending to be drowned in a sea of wonder, yet if asked what they are wondering at they do not know. "They should be told to wonder as much as they please, but at the same time to remember that the Almighty is their Creator and that they are His servants."

This rebuke by Al Ghazali echoes our attitude regarding the fallacy of the popularizers. They are in awe of the natural universe yet deny the reality of the Nothing that lends reality to the natural universe.

Finally, we also see that the *via negativa* has been very much alive in the history of mainstream Rabbinic Judaism. It has not been left to mystics alone to deny the efficacy of the human intellect. For example, Moses Maimonides, the renowned Rabbi of twelfth-century Spain composed his *Guide to the Perplexed* mainly to argue against the trend of anthropomorphizing God. (Many Jewish people know Maimonides by the acronym Rambam—*Rabbeynu Mosheh Ben Maimon*—he would be very cool on urban streets in the West today!) He states his apophasis by warning that describing God by negation is the most penetrating way and implies no deficiency in God. "On the other hand, if one describes Him by means of affirmations, one implies... that He is associated with that which is not He and implies a deficiency in Him."

Let us now conclude this theme on Nothing and nothingness by adding three more interesting teachings from the Kabbalah, a widespread mysticism born out of Judaism in the thirteenth century.

Kabbalists were so aware of what Nihilology terms the Nothing and nothingness of the divine that they envisioned Yahweh with two distinct natures: one is essential, transcendent, and ungraspable, while the other is manifest, personal, and revealed to human beings. This first nature is often called *Ein Sof,* "that without limits." *Ein Sof* refers to Yahweh before self-manifestation through creation of the physical world. (Intriguing to note that the term we write as 'Yahweh' was actually four consonants without vowels in biblical Hebrew, i.e., a name for the incomprehensible god that is unpronounceable.)

Another term used by Kabbalists in teachings about creation is *ayin,* "the nothing." *Ayin* describes the nature of the material world before creation. We see it used in opposition to the term *yesh,* "the something." *Yesh* describes the material world in manifest form. Now, for our task to build the heuristic Nihilology, it is important to note that *Ein Sof* and *ayin* are so closely akin in Kabbalah that they often overlap in meaning.

One of the root texts of Kabbalah was composed in thirteenth-century Spain by the mystic Joseph Gikatilla. In *Sha'are Orah, The Gates of Light,* he gives teachings on the 300 names of the divine. He held that from these names flowed a fountain of living waters to quench the spiritual thirst of human beings. One of these names is *ayin*. "The depth of primordial being... is called ayin... If one asks, 'What is It?' the answer is, 'Ayin,' that is, no one can understand anything about It... It is negated of every conception."

Daniel Matt, the modern-day Jewish theologian states in his work *The Essential Kabbalah: The Heart of Jewish Mysticism* that the word *ayin* carries connotations of negativity and non-being, but what the mystic means by *ayin* is that God is greater than any thing imaginable; it is no thing. "Since God's being is incomprehensible and ineffable, the least offensive and most accurate description one can offer is, paradoxically, nothing."

Here we see again from Apophatic Theology a confluence of notions regarding Nothing that aids us in building Nihilology: *Ein sof*, the Nothing of Yaheh is akin to the term *ayin*, the Nothing of materiality. In a sense, they are differentiated by the context of the theological discussion at hand. When reflecting on what lies beyond the continuum of materiality, the Kabbalist often uses the term *ayin*. When reflecting on what lies beyond the continuum of divine personhood, the Kabbalist often uses the term *Ein Sof*. But sometimes the terms are interchangeable. Material Nothing is equivalent to Yaheh Nothing.

Now, let us go one step further in our gleaning of teachings about Nothing and nothingness from the Kabbalists. They state that the essential nature of Yahweh is Nothing (*ayin*), yet this Nothing is more real than creation (*yesh*). For example, David ben Abraham ha-Lavan, a fourteenth century Kabbalah leader taught that Nothing [*ayin*] is more existent than all the being [*yesh*] of the world. This position is echoed by Meister Eckhart, a contemporary Christian mystic. "God's Nichts [nothing] fills the entire world; His something, though, is nowhere."

Thus, according to the Kabbalists—and this teaching is in complete accord with the Buddhist and Taoist teaching discussed before—Yaheh permeates the entire created world as Nothing.

And finally, it is important to briefly touch on the teachings of Kabbalah regarding the possibility of Yahweh as *ayin* to be known. This short mention will serve as a foreshadow of the elaborate discussion in the section below called Nihilitive Knowing.

First, let us note the words of the fourteenth-century Spanish mystic David ben Judah he-Hasid from his oft-read text *The Book of Mirrors*, where he asks and answers the essential epistemic question regarding this reality beyond form or *ayin*. Can a human being actually know *ayin*? "However, concerning the cause of causes [*ayin*], there is no aspect anywhere to search or probe, nothing can be known of it... therefore forgetting pertains to the comprehension of this place."

Also, let us add the perspective of Issachar Ber of Zlotchov, echoing the teachings of *Zohar*, the most important Kabbalah text of all. He urges humans to turn away from objects of thought toward the place called nothing (*ayin*) and says this is the essential act of worshiping God. "Then one comes to the state of *ayin*... One has no independent self, and is contained, as it were, in the creator."

The ability of human beings to actually know the divine as Nothing most likely strikes our readers as an obvious impossibility. This possibility is boldly and clearly averred to be the case by the leading voices of Kabbalah. However, they also call upon the linguistic dance of paradoxes to make their avowal: knowing by unknowing, knowing by forgetting, thought without object of thought. Below we will make clear sense of this epistemological conundrum and will not succumb to the language of paradox.

It should, in addition, be noted that our discussion of Apophatic Theology in Judaism and Islam, when taken as a whole, has brought to light the confluence of three notions regarding Nothing: the Nothing of God, the Nothing of the world before creation, and the Nothing of the human in a state of knowing God. ("One has no independent self, and is contained... in the creator.") This confluence is an important perspective for the building of Nihilology and for its effectiveness as a heuristic tool. It is a shocking confluence, to say the least: The underlying nature of the divine, the manifest world, and the human being are one and the same!

One more confluence of notions should be noted in concluding this theme. A disciple of Issachar Ber of Zlotchov say, "When one attains the level of gazing at *ayin*, one's intellect is annihilated. Afterward, when one returns to the intellect, it is filled with emanation."

When a believer contemplates the divine as Nothing and comes into the presence of the divine as Nothing, their intellect is "annihilated." Yet, when the intellect comes back from its silencing, it will be enhanced with light. This statement from Kabbalah bolsters our interpretation of outsight as introduced above in our discussion of scientific discovery. Again, we see the confluence of notions. "Knowing" God as Nothing comes about with the silencing of intellect. When intellect revives, it is filled with light; this light, in our view, could manifest in the realm of thinking as outsight.

The Manifest World?

We have hinted and foretold many times thus far in our book that we will explain and give a creative interpretation of how the divine may be encountered, what we term Nihilitive Knowing. And we will do so, yet first it is important to delve into two more themes in order to lay groundwork for that discussion. Before we can get a true understanding of how humans know divine Nothing, we need to understand how we know the manifest world as something. We also need to know what is wrong with that something.

Let us begin once again in the East with *Advaita Vedanta*, a school within the wider religion of Hinduism. Of all the teachings that have been given voice in the theological history of the Indian subcontinent, *Advaita* is the most clear in its expression of (or obsession with) probing the true nature of the manifest world.

The most revered spokesperson of *Advaita* is the eighth century wandering monk Shankara. His period in Indian culture was one of rampant distrust in the common daily life. Thus, as a very young man, Shankara took vows that announced to his family and all those around him that he renounced all roles incumbent upon him in society.

Before extending our discussion into the nature of the world according to Vedanta, let us note that all forms of theological thought, East or West, come to be in a particular place and time. This context for ontology must be regarded in order to deeply appreciate the meaning of those theologies. We call this context the birthing mud. The theology of Vedanta was born in a time when people were disgusted with worldly life and could find no meaning in things.

Further, this birthing mud gave forth a sort of behavioral, attitudinal response and strategy that encouraged one to renounce the world, to assume a discipline that sought release from all life in the world. Fulfilling the duties of father, householder, businessman, or priest was ultimately a means of accruing sticky karma that would weigh one down and guarantee another life on earth in the future. The sought-after goal of the renunciate religious life was to rid oneself of being born into the world at all. Advaitans called this goal *moksha,* release from rounds of birth and death.

The Brihadaranyaka Upanishad, one of many canonical texts upon which Shankara based his teachings, says about the ultimate reality of Vedanta that it terms *Brahman*, "Now, therefore, the description [of *Brahman*]: *Neti, neti*——'Not this, not this.' Because there is no other appropriate description than 'not this.'"

"*Neti neti*" is the Advaita form of Apophatic Theology. "Not this, not this" rids the renunciate of self attributes and *Brahman* of divine attributes to clear the path to non-differentiation (Thus the name of the school: *Advaita*, or non-dual). David Paul Boaz, in his book *Shankara's Advaita Vedanta*, helps us understand *Brahman*. "One, without a second, without limit, empty of all predicates, attributes and qualities, beyond concept and belief, or any subject-object dualism whatsoever." Shankara calls this unity by the traditional name *Nirguna Brahman*. This *Nirguna Brahman*, the "ultimate without qualities," is not a personal deity at all, rather an impersonal absolute. The Sanskrit word finds its roots in "expansion" and implies here that *Brahman* is infinite expansion, infinity. This ultimate is what Nihilology refers to as Nothing.

Now, Vedanta taught that the absolute *Brahman* does not manifest into the world. Rather, the world is taught to be *avivarta*, a superimposition on *Brahman*. The root term in Sanskrit comes from *vrt*, to roll. So *a-vivarta* implies that the world has never rolled out at all. Indeed, the distinction of world from *Brahman* has never factually occurred. The world is not a something! As Shankara states in verse 252 of his pinnacle opus *The Crest Jewel of Discrimination*, "As dream-built lands and times... are all unreal just so here in waking is this world... in as much as all this world... [is] unreal."

The manifest world of Vedanta is what we call in the realm of science a virtual reality. VR scientists employ computer simulations to allow a person to interact within a synthetic three-dimensional visual and sensory space. By using interactive devices that send and receive information, such as goggles, headsets, gloves, or body suits a person becomes immersed in a computer-generated space. These spaces are extremely engaging, and with long immersions, a person becomes convinced of the genuineness of the space. The Vendantist teaches that it is exactly the same with the ostensibly manifest world. With life-long immersion, human beings come to believe in its reality, when in fact it is almost real, the manifest world is virtual.

It is also interesting to note that VR science has borrowed the term "avatar" from the Sanskrit lexicons of Vedanta and other Hindu religious traditions. The avatar is the persona within the virtual space of the person using the goggles or body suit. "Avatar" is the modern-day pronunciation of *avatara*, one who "comes down" in to the virtual manifest world from the absolute realm to aid human beings in extracting themselves from their immersion in the almost real world.

Vedanta calls the manifest world *vyavahara*. It is delineated, and those delineations are given names. It is an empirical reality, a space of common experience and of practical activities. In this realm the things of the world are experienced as real. In the ontology of Vedanta, the world is not totally unreal, like a horn on a hare, as their teachers like to say. Nor is the world totally real, like *Nirguna Brahman* which is the only real. Rather, the world is real like a dream. It is experienced by the dreamer, yet it disappears upon awakening.

To go one step farther, in this *vyavahara* world, the individual soul and the worshiped personal deity also are experienced as real. As we have seen, *Nirguna Brahman* is ultimate reality without qualities. *Saguna Brahman* is the personal god with qualities that virtually comes to be in the *vyavahara* world. This divine person is often called *Satchitananda*, having the qualities of truth, consciousness, and bliss. In the language of Nihilology, Nirguna Brahman is Nothing, having the nature of "not this, not this" and about which no true statement can be made. *Saguna Brahman*, the personal ultimate with qualities, is nothingness, he about whom one can form concepts.

Yet, what is it that causes this virtual space of *vyavahara* to delineate? Shankara says this world "is *maya*, whereby all this moving world is made to grow." This term *maya* finds its root meaning in "wisdom to create," and as such, is a name that gives title to the creative force in the universe. Yet, *maya* is also the power to create magic. Often this magic power is personified as a female deity who confuses humans with her dance. And this is the intent in *Advaita Vedanta*: the force of world creativity is an illusion maker, tricking humans into experiencing the world as having ontological status as separate from *Brahman*.

Thus, for Vedanta, the world only seems to be differentiated from *Brahman*, the absolute without qualities. Human beings fall for the magic dance of *maya* due to what Vedanta sees as the inbuilt flaw of the world, human ignorance. (We will discuss this basic flaw as witnessed in the religious traditions of the East and West below.) Ignorance has us superimpose concepts and names and qualities on *Brahman*, thus presenting us with an empirical world that we take as real. It seems that Shankara would conclude that the popularizer physicists and cosmologists of today are ignorant.

Let us now turn to Samkhya, another school within the Hindu tradition that arises from the same birthing mud as Advaita (non-dual) Vedanta, yet posits a dualistic universe. As Rodney wrote in his book *The Samkhya Path of Observation*, when the Samkhya seeker observes his world, he sees an utterly lopsided duality. On the one hand there is *Prakriti*, the entire existent cosmos, in her manifest and unmanifest forms. On the other hand, there is *Purusha*, the individual knower of this world. The duality of the Samkhya world is absolute. Both aspects are eternal and eternally real.

This dualism of Samkhya is a very lopsided one. On the one hand there is *Purusha* ('human'), as the fact of consciousness, and *Prakriti* ('producer') as the fact of existence. The essence of the human being is consciousness, everything else is *Prakriti*, who in her unmanifest form (*avyakta*), causes the manifest world (*vyakta*) to come about. The created world is also *Prakriti* herself; thus creation is a going forth of *Prakriti* into manifestation. Samkhya posits a universe that comes to be out of an eternal existent.

The verses of the Samkhyakarika, the urtext of the school, express this lopsided universe in that they are almost solely—90% of the verses—concerned with *Prakriti*. Though the essential reality of the human being is *Purusha* as consciousness, though the striving of the Samkhya seeker is for the sake of realizing this reality, the verses of this central text are about what the human being is not. Samkhya, like Advaita Vedanta discussed before, shows an almost obsessive interest in the workings of the manifest world.

We stated earlier that the force of world illusion, in Vedanta called *maya*, is sometimes portrayed as a female who mystifies human beings with her dance. This is even more so the case with Samkhya's *Prakriti*, where the "it" of the creative force is generally anthropomorphized into a dancing "she." And *Prakriti's* dance has no beginning in time, although she does rest now and again. These rest periods are eon-long periods of dissolution. The ever-recurring cycles of going out into manifest form and dissolving into unmanifest form is the entire performance of *Prakriti.*

Prakriti is said to be utterly constituted of qualities called *gunas.* They are three in number: lightness, heaviness, and movement. The *gunas* function much like the yin and yang of Taoism or like the four elements—air, earth, fire, and water— in the medieval annals of Alchemy. They exist to provide an analysis of the basic makeup of the world. And when *Prakriti* does take a rest, that is, when the world dissolves into the unmanifest state, the qualities that make up her totality continue to exist.

A parallel notion from modern creation physics may help us come to terms with this rather paradoxical teaching of Samkhya that *Prakriti,* even when unmanifest, is still comprised of productive qualities. Many cosmologists maintain that the world came to be, naturally and of its own accord, out of a quantum vacuum. However, according to *Scientific American* author Philip Yam, "Energy in the vacuum, though, is very much real. According to modern physics, a vacuum... churns with unseen activity even at absolute zero, the temperature defined as the point at which all molecular motion ceases."

Stephen Hawking, father of the big bang, held that the quantum vacuum is inherently unstable. The energetics in this pre-manifest state cause the vacuum to fluctuate, periodically and spontaneously. Many physicists argue that the universe itself is the result of such a quantum fluctuation, giving forth so-called virtual particles that found a stable life as matter. Stephen Battersby puts it this way in a lighthearted flourish of scientific writing: ", "Matter is built on flaky foundations.

Physicists have now confirmed that the apparently substantial stuff is actually no more than fluctuations in the quantum vacuum."

Applying this scientific language to Samkhya, *Prakriti,* in her unmanifest state *(avyakta)* inherently comprises of unstable productive energies (*gunas*) and causes the world to manifest *(vyakta)* out of her self. In the language of Nihilology, the Samkhya world is a something manifested from a pre-existent latency.

Now, please allow us one philosophical flourish of our own regarding language that depicts Nihilology's challenge in elucidating Nothing. Take a look at the words theist and atheist, manifest and unmanifest, *vyakta* and *avyakta*. As human beings of worldly experience, we define things positively in terms of what we know first and from there find the domain of the **un**-known. Isn't it quite revealing that we **add** to words that give name to our experience the prefixes a-, un-, and non- to signal a negation or an undermining of what empirically is. Of course, in the logic of linguistics, this is done because it is a systematic way of denoting opposites or negations, but might this same tendency contribute to why Nothing eludes us? The more we expand our knowledge (and the more we realize how much remains unknown) requires us to **add** to conscious understanding, never to **subtract** from it. We might conclude that we are destined to follow a linguistic version of the second law of thermodynamics that determines a unidirectional, irreversible trajectory away from Nothing. Thus, could it be so, that the more we attempt to order our thoughts (through addition), the further we push ourselves from our initial state (subtraction unto Nothing)?

Returning once again to the teachings of Samkhya on world creation, we see that they are seemingly oppositional to those of Vedanta. Samkhya posits a cosmic duality of reals—consciousness and world, whereas Vedanta holds to a monism—*Nirguna Brahman* overlaid by a dream. However, with a longer steeping in the Hindu world of these classical times (Remember Vedanta and Samkhya both arise from the same birthing mud) we realize that these two schools are united in one overarching attitude toward the manifest world: it comes to be due to human ignorance (*avidya*).

Ignorance, according to Samkhya, consists of a false identification of *Purusha* with *Prakriti*, a false identification of consciousness with the activity of the world. Both are real, but ignorance has us identified with the wrong real. In the words of verse twenty of the Samkhyakarika, "By means of their relationship the inconscient one appears as if characterized by consciousness; likewise the indifferent one appears to be an actor, even though the agent of activity is the gunas."

The ignorant human experiences life thus, "That which is me is conscious. That which is me has desire. That which is me will take action." False identity, ignorance, means uniting *Purusha* and *Prakriti*, making them both "me." Setting oneself free of ignorance demands differentiating the reals, i.e., seeing that I am *Purusha* as consciousness only. All else is Prakriti, not me.

Samkhya tells us that *Prakriti* begins to dance in order for *Purusha* to observe her. In other words, the world comes to be in order for to us see it, and in this seeing our misidentification is overcome. As verse twenty-one of the Samkhyakarika states, "The relationship of those two, like the blind and the lame, is for the purpose of *Purusha*'s seeing [*Prakriti*], and for his spiritual completion. Creation comes forth out of that relationship."

This verse answers the question, "Why do I have manifest experience at all?" As repeated many times in the Samkhyakarika, the manifest world is *Purushartha*, for the sake of *Purusha*. The manifest world gives

Purusha exactly what he needs to see in order to free himself from ignorant false identification with the world, i.e., from experiencing body, thoughts, and possessions as "me." In the anthropomorphic language of Samkhya, *Purusha* choreographs *Prakriti*'s dance simply by sitting quietly in the audience. Whether she portrays Shiva as the cosmic dancer or earth-shaking lover depends on what *Purusha* needs to see. Or, in a more theological analysis, the *gunas* express themselves in specific arrangements which result in certain physical bodies, subtle bodies, and certain life situations, to benefit *Purusha* in his search for spiritual wholeness and perfection. Therefore, not only does needing to watch *Purusha* stimulate the world coming into form, it also determines the specific form the world will take. In its most astounding teleological statement, Samkhya says that the world is there only for us and in the exact form we need to find our independent integrity once again.

Once again, it may help us to understand this Samkhya teaching as to the effect of observing consciousness on the manifest world by referring to a similar tenet of physics. Before we discussed the famous double slit experiments that showed if an observing mechanism is set up to see which slit the electrons actually pass through, the exiting electrons no longer act as waves, as they had before. The very act of watching the behavior of electrons alters the nature of the electrons themselves.

In classic science, the ideal observer is one who doesn't perturb the system being observed in any way. This is the basis of the objectivity that science has been so proud of. Yet in the realm of quantum mechanics, as in the double slit experiments, the observer and the system being observed are intimately and mysteriously related, so that the results of any observation are nuanced by the observer.

Referring here to the paradox of affecting a particle through measuring its locality, Marcelo Gleiser boldly states in his book *The Island of*

Knowledge: The Limits of Science and the Search for Meaning that when some form of measurement takes place the particle ceases to be a potentiality and becomes a reality. "It does so instantaneously, shrinking from being everywhere it can be to being localized in a small region of space.... More dramatically stated: to measure is to create.

Many physicists have extended this notion of measurement to assert that the consciousness of the observer is an integral aspect of how creation comes about. This position echoes Samkhya. Indeed, the Samkhya teachers were quite good quantum mechanics!

To come home to the West now, it is clear that arguments about whether the world came to be out of Nothing (Vedanta) or out of a pre-existent stuff (Samkhya) can be alive within a single tradition as well. For example, let's take the case of the early history of Christianity.

We generally take for granted that Christians believe the world was created on the first sacred day *ex nihilo*. This is a Latin phrase that translates as "out of nothing." The actual phrase *ex nihilo*, of course does not appear in the early verses of the biblical book of Genesis in that it was orally passed on then written down in Hebrew. Here are the relevant creation stanzas according to the *New Revised Standard Version Bible with Apocrypha*:

In the beginning when God created the heavens and the earth,
the earth was a formless void [*tohu va vohu*]
and darkness covered the face of the deep
while a wind from God swept over the face of the waters

The phrase *tohu va vohu*, as we discussed before, denotes a formless stuff. And in Genesis it is associated with water in that it is uniform and unstructured. The phrase **connotes** a chaos, uncontrolled and dangerous. The most important point for us to understand is that *tohu va vohu* is not Nothing, not a *nihil.* The authors of the New Testament give witness to this earliest understanding of these scriptural words about creation. For example, the Apostle Peter wrote regarding "scoff-ers" in his second epistle. "They deliberately ignore this fact, that by the word of God heavens existed long ago and an earth was formed out of water and by means of water." Peter, in his "of water and by water," emphasizes that the world came to be of and by a prior chaos.

New Testament voices who were familiar with the Hebrew texts of Genesis took as revealed truth that the world came to be out of an eternal something. Even later into the first and second centuries, many church fathers, such as Theophilus and Origen, held this view in their theological writings. As an example of how undiscriminating—and uninterested—the early church was about the specific ontology of creation, it is well known that the great Justin Martyr accepted in a simple faith manner that Adam was created out of the earth, as witnessed in the biblical account of Adam and Eve's ideal life in the Garden of Eden.

It was not until the late second and early third century that the phrase *ex nihilo* took on a formal ontological meaning in the Latin church. And this formalization, that is, making into orthodoxy, was perceived as necessary in order to quell the fear of an ascent of a growing schismatic force in the church called Gnosticism.

There were three central tenets taught by the Gnostics that the mainline Latin Church found abhorrent and heretical (In addition to the fact that Gnosticism was a phenomenon of the eastern church that was experienced as less civilized than the church in Rome). First, the world was not created directly by God, rather through the labors of a lesser divine known as the Demiurge, a sort of craftsman who actually formed up the eternal stuff. Second, the physical world held a taint for the Gnostics; it was inherently evil and to be avoided by the truly religious. Whereas orthodoxy held, as witnessed in Genesis, the earth was "good" and holy in that Christ was incarnated there. And finally, Gnostics taught that the Demiurge created the world out of a chaotic, eternal stuff. Eternal stuff allowed for a real existent alongside God and thus his uniqueness is threatened. God, according to the orthodox view, must have created from nothing.

In this intra-familial battle during the late second and early third century, Iranaeus stands out as the most devastating Latin church apologist. (What a tender title for a staunch defender of mainline theology!) In his important and effective work *Against Heresies*, he argues that the Gnostics don't have the capacity to appreciate the miracles God brings forth. "While men, indeed, cannot make anything out of nothing, but only out of matter already existing, yet God is in this point preeminently superior to men, that He Himself called into being the substance of His creation, when previously it had no existence."

Orthodoxy, indeed, won out and squelched the Gnostic movement. By the sixth century it was all but extinguished. The Latin Church even did a thorough job of erasing the history of Gnosticism's texts, churches, and practices. They were no longer a threat.

It is interesting and revealing to note that with the death of the Gnostics movement and with it the ongoing need to apologize for the unique nature of God, the ancient stuff has once again returned to modern Christianity as a popular mythic—not theological/ontological—worldview. Karen Armstrong says in her book *In the Beginning*, "God did not make the world out of nothing, since the waste of chaos and the primal sea were already in existence, and God merely imposed order upon the *tohu va-vohu...*"

It should be noted, in order to maintain clarity in our overall exploration of religions regarding Nothing and nothingness toward building the heuristic of Nihilology, that "nothing" in the context of third century Christian interpretations of creation means a lack of any substance, water, chaos, or stuff out of which the material world was made manifest. (Recall from our early discussions that Nothing lies off the rarefied end of the continuum of materiality.) Christian Fathers were not expanding their *nihilo* to the nature of God himself. History had to wait until the second millennium of the common era for that. However, this expansion is precisely what the Jewish mystics of Kabbalah did as much as a thousand years earlier.

We discussed ealier the confluence of notions regarding Nothing among the Kabbalists. Yet, just allow two more quotes to bring home the point. For the Kabbalist, the deepest mystery is the transition from *ayin* to *yesh*, from "nothing" to "something." Daniel C. Matt says, "There is a 'something' that emerges from 'nothing,' but the nothing is brimming with overwhelming divine reality. The something is not a physical object but rather the first ray of divine wisdom, which, as Job indicates, comes into being out of *ayin*."

This second statement is a more traditional one from Rabbi Simeon ben Jochai, the second-century miracle worker rabbi, who avers that God as *ayin* comes forth as the primary element of the world. "...the illimitable phosphorescent ether, of the nature of light, formless, colorless, being neither black nor green nor red. In it, latent yet potentially as in a mighty womb, lay the myriad prototypes and numberless forms of all created things."

These statements together re-affirm our conclusions before, that the Jewish mystics of Kabbalah brought the material Nothing out of which

God created and the divine Nothing who created the world into such close kinship that they became ontologically identical!

Now, to conclude this theme, there is a clear difference between East and West in the religions' attitude toward the manifest world. In the East there has been an obsession with analysis of the manifest world and how it can be known. This has not been the case in the West where the obsession has been more with the analysis of the divine and its/his relationship to humans or human culture. This East to West contrast has also been stark regarding the inherent reality or virtual reality of the world. This contrast will be revealing later as we critique of the popularizers of science.

The Basic Flaw: Something is Wrong in the World

All major traditions of the world, East and West, teach as a grounding tenet that there is something wrong with manifest existence. This wrongness lies with humanity, and it is the *raison d'etre* of religious life to overcome this wrongness and lead humans to a perfected life. What is wrong may be seen as ignorance or moral fallenness, yet some flaw runs deep in the human being. We might say that religions approach this flaw as the defining characteristic of the human condition.

Teachings on the range of human ignorance in the history of Buddhism evolved and universalized over a period of five or six centuries. In the beginning, ignorance was about the self. The great philosophizer of the Mahayana School, Shantideva, reflected on Buddha's original words in these verses from his collection *The Way of the Bodhisattva:*

> For beings long to free themselves of misery.

But misery itself they follow and pursue.
They long for joy, but in their ignorance destroy it
as they would an enemy

How much suffering and fear
How many harmful things are in existence
If all arises from clinging to the "I"
what should I do with this great demon?

A modern-day Mahayana Buddhist from Tibet, Ringu Tulku, in his book *Daring Steps Toward Fearlessness*, brings the words of Shantideva into modern western parlance in stating that ignorance is at base the false identification of self as independent. "It consists of the belief that there is an "I" that is not part of anything else. On this basis we think, 'I am one and unique. Everything else is not me.'"

The Buddha's great awakening led him to teach that the craving, thirst, and suffering of human life finds its root in the false experience of individuality. In truth, he taught, there is no unique self. As Buddhism evolved, the range of ignorance expanded to the absolute limit, as we saw in our earlier discussions; all the manifest world, self and things, are empty of unique being. Only in the context of the eternal Nothing (*shunyata*) do people and things find reality. With the realization of the presence of Nothing in all things, the human being no longer suffers.

The flaw in the universe is also taught to be human ignorance in the Hindu traditions, as witnessed in Advaita Vedanta and Samkhya. For Vedanta, the locus of ignorance is the human mind. As discussed earlier, the central teacher of Vedanta was Shankara. In his crowning work *The Crest Jewel of Discrimination*, he gives forth frilly and precise poesy:

For there is no unwisdom, except in the mind

for the mind is unwisdom, the cause of bondage to life...
In dream, devoid of substance, it emanates a world...
which is all mind.

Man's circle of birth and death comes through attributing reality to the unreal
but this false attribution is built up by the mind
This is the effective cause of birth and death and sorrow... "

Under the sway of mental ignorance, human beings mislead themselves
to locate their self in their body, thoughts, and individual life events
Overcoming this falsity leads to the true experience of self as the one and only real
Brahman without qualities, and again all suffering ceases.

This [true self] is not born nor dies nor grows
not does it fade nor change forever; even when this form [body] melts away
it no more melts than the air in a jar [when the clay is shattered]

Where the cause of delusion melts away
like darkness in light
in the secondless supreme reality, undifferentiated
what separateness can there be?

Now, the basic wrongness in the world, according to Samkhya, is also a form of ignorance. Yet their teachings speak of a false identification of human consciousness (*Purusha*) with the created things of the world (*Prakriti*). They are both real and eternal, yet we identify them both as "me." Verse forty-eight of the Samkhyakarika emphasizes the myriad ways in which a human being can reap the rewards of being ignorant. "There are eight varieties of obscurity and delusion, ten varieties of extreme delusion. Both gloom and utter darkness are eighteenfold." Samkhya paints a human life as one of extreme, and multi-faceted, suffering when ignorant of our self as knowing consciousness. To differentiate between what is me and what is not me is the way to end suffering.

Echoing verse two of the Samkhyakarika, Gerald Larson in his foundational book *Classical Samkhya* says that common human life on earth is by its very nature one of suffering that cannot be removed by means of drugs, medicine, or scripture. "Only discriminative knowledge (*vijnana*) is effective as a means and specifically knowledge of *avyakta* (the unmanifest), *vyakta* (the manifest), and *jna* (the knower); i.e., of *Prakriti*, of the manifest world, and of *Purusha* respectively."

Turning back to the West now, we see that Christianity teaches a basic human flaw as well, yet this tradition's flaw is not one of ignorance but of disobedience and haughtiness. Humans are said to be in a state of sin, guilty and missing the mark. In his book *The Flaw in the Universe: Natural Disaster and Human Sin*, Adrian Hough tells us that "sin is the dark side of the human being. It is the_basic human state or condition which disrupts the relationships of which we are a part and which therefore results in the possibility of our committing... moral misdeeds in the first place..."

The gospel letters of the New Testament tell us that we inherited our condition from Adam's deeds in the Garden of Eden. The book of Romans describes our ancestral estate:

Therefore... sin came into the world through one man
and death came through sin
and so death spread to all because all have sinned

Just as by the one man's disobedience
the many were made sinners
so by the one man's obedience
the many will be made righteous.

Death awaits all of us because of our inbuilt condition of disobedience and haughtiness. And in a fascinating theological turn, we as human beings are incapable of working our way out of this condition. Christianity teaches that it is only through the grace of the incarnation that sin is overcome. The sense is that human beings who strive to free themselves from sin are deluded by the very haughtiness they are trying to overcome!

Now, for our work it is essential to detail just how, according to Christian teachings, our sin came about. The third chapter of Genesis records a conversation between Eve, the first woman, and Satan, the personification (or reptilization) of sin. Satan tempts Eve, and thus Adam as well, to partake of a certain tree in the garden that God has forbidden Adam and Eve to eat, "For God knows that when you eat of it, your eyes will be opened and you will be like God, knowing good and evil."

The phrase "good and evil" translates the Hebrew *tov V'ra*. Biblical scholars tell us that *tov V'ra* functions here linguistically as a merism, a figure of speech by which a totality is denoted by referring to some of its constituent parts. Again, as we discussed earlier, it is notable how Genesis contends with concepts that induce fear by calling on playful, lighthearted language. *Tov V'ra* is akin to saying "hook, line, and sinker" (the whole subject) or "sun, sea, and sand" (the entirety of a Caribbean vacation). Such merisms were put to use throughout the ancient and classical eras of Egypt and Greece. They are seen elsewhere in the Hebrew Bible as well. For example, when Genesis states that God created the "heavens and the earth," this refers to the coming to be of the entire universe. Or in Psalm 139 it is taught that God knows "my sitting down and my rising up." This is an everyday way for the psalmist to aver that God knows every deed he performs.

So, the sin of Adam and Eve that we have inherited through the history of mankind is one of seeking to know everything, suffering under the delusion that we can actually achieve omniscience.

It may assist our understanding of this biblical teaching on the fatal sin of disobedience and haughtiness to explore it in terms of modern western psychology, namely in the work of Carl Jung who had as one of his life goals reconciling the realms of science and religion.

Edward Edinger in his classic work *Ego and Archetype* echoes the words of Jung. "Modern people assume there is nothing that they have not made up.... our God-all mightiness, which simply comes from the fact that we think we have created everything psychical—that nothing would be done if we did not do I."

According to Jung, this great hubris of human beings, this "God-all mightiness" is what strikes us with a sense of loneliness in our own minds, because we hold to the notion that the extent of the mind is determined by our self-originated thoughts and perceptions. Jung recommends the remedy of emptying, an internal psychological process of cleansing the ego tendencies that blind a person from their deeper relationship to the Self. The Self is the encompassing identity of the human being that finds its locus within what Jung calls the collective unconscious, the transpersonal psyche that all humans share.

(When we turn our Nihilology to a critique of the popularizers, we will see that they, in Jungian terms, are guilty of separating their ego from the collective unconscious without returning the debt they owe to the Self. We may be prone to a bit of a reminder in their own terminology: they are guilty of breaking the law of conservation of energy where virtual particles pop into existence but must quickly annihilate to maintain the energetic balance of the universe.)

This Jungian interpretation of the nature of ego and Self can act as a helpful guide to understand the why behind the Christian mythos. In Christian teachings, this ego emptying is called *kenosis.*

In the Gospel of Matthew, for example, Jesus preaches his pinnacle Sermon on the Mount. In it he declares what are known as the beatitudes, eight pithy poems that give praise to the non-inflated person, one who empties themselves, and in doing so, the sacred Holy Spirit can enter. Then the person becomes who they truly are.

There is also, of course, the famous statement by St. John of the Cross in *The Dark Night of the Soul.* "It is meet then that the soul first be brought into emptiness and poverty of spirit and purged from all help, consolation and natural apprehension with respect to all things, both above and below."

Thus, this Christian *kenosis* teaches that the person is brought to the point of deflation, emptied of hubris, so the Self can enter.

And so, to encapsulate this whole theme on the basic flaw in the world, let us foreshadow once again our use of the teachings of the world's religions to critique leading modern scientific thinkers. These popularizers see the human being as a simple occurrence in the universe, having no special status therein. The human carries no special significance in the grand cosmic scheme of things. This position connotes a happenstantial quality for humanity, one among an infinity of creations.

This stance has allowed for great scientific discoveries such as the heliocentric galaxy, creation as many more than 4,400 years ago, and the evolution of species. However, in relegating the human to such a status, they have fallen victim to three basic flaws of the universe, the wrongness of creation, as taught by religious traditions. First, they experience themselves and the world separated from Nothing, taught by Buddhism and Taoism to be ignorance. Second, they identify their body and mind as the extent of who they are and so are not aware of their selfhood as something other, Vedanta and Samkhya's form of ignorance. Third, as taught by Christianity, they do not consider themselves as the heirs to Adam's sin and thus do not see that seeking the knowledge of everything is indeed hubris that leads to death. The popularizers have fallen victim to all the flaws of the manifest world, as taught by the major religions of the world.

(4) Nihilitive Knowing

Topology of Knowing

One way to proceed here, one that goes a long way toward appreciating what we mean by Nihilology, is to highlight a single thread of our discussion thus far. It has not been hidden, yet may not have etched deeply in the mind's eye of our readers because there has been such an array of bright colors going into the fabric of our argument thus far. This thread may be referred to as ways of knowing, what philosophers call epistemic pathways.

In order to highlight this thread, we will put forth a structured analysis of epistemic pathways that are witnessed in the teachings of world religions. We will focus on two most essential to our building of Nihilology: Faith as heuristic knowing, and Nihilitive knowing. There will be, of necessity, much harkening back to earlier sections in order to weave this discussion together. In this, it is our goal to bring to culmination much of what we have intimated in the previous section regarding Nihilology as a heuristic for understanding Nothing and nothingness.

To begin, it is helpful to ask this simple question: Where does knowing reside? There is a three-part typology that philosophers of knowledge often employ to analyze sorts of knowing: Objective, Subjective, and Intersubjective. We approach these sorts of knowing from a decidedly spatial perspective. Thus, let us call it a three-part **topology** and proceed with our analysis on the basis of "where."

Objective truth is invariant; it does not evolve or devolve over time, even as the holder of the truth matures or restricts their view of a matter at hand. In the language of epistemology, objective truth can be stated in the form, "I understand that..." This stated form is called a proposition, something that can be expressed in an expository manner purporting to describe a fact or a state of reality, such as "Whales are mammals," "56 x 19 = 1064," or "$e = mc^2$." Objective truths are, as we often have discussed, the pursuit of scientists. Yet any person, at whatever place and time, will judge these truths correctly if they judge according to the procedures of the scientific method.

One of the foundations of the scientific method is that all objective propositions must be available to verification via experimental replication and expert critique by peers employing the correct procedures of rational analysis. The distinctive claim of science is that propositions which stand up to these tests of verification are, as such, valid.

So, where does objective knowing reside? It lies in an individual as they look outward to the world around them. It resides in the mind regarding that which is objective to the individual knower, i.e., outside their body. It is also objective in a third sense. Verification of objective knowing, the process by which it achieves the status of truth, occurs in the outside world.

Now, subjective knowing gives birth to a truth that is unique and particular to the individual person. And, unlike objective knowing, it is open to change as the individual may mature or alter their attitudes in another way. Subjective truth is also not verifiable in the objective manner of science.

Subjective knowing often concerns internal states of experience, "I am too hot here on the beach," or "My mind is just too cluttered today to complete anything worthwhile." In the language of epistemology (distinct from objective knowing's "I understand that..."), subjective knowing is expressed in the form, "I feel that..."

Philosophers of mind and cognitive psychologists have discovered that what one feels is integrally related to what one experiences as known. This union of emotion with knowing they term epistemic feelings. And, according to the research of cognitive scientist Santiago Arango-Muñoz, they are very physical in nature. They are embodied! Epistemic feelings are based in a physical condition occurring internal to oneself and give very specific witness to those bodily sensations.

Muñoz also points out that epistemic feelings are essential to many forms of subjective knowing, such as remembering, planning, and imagining. Further, even though subjective knowing is inward and individual, this inner experience may reference a thing or event in the outer world. In this particular function, subjective knowing is often referred to as "gut feeling."

A gut feeling of a thing is shaped by how a person experiences that thing in the outside world. (Let us ignore the subtleties of metaphysical idealism. For now, we will accept that real things exist.) One could call this gut feeling an encrypted catalog of such experiences, a sort of mess set of relations of exemplars between themselves, the world, and oneself. For example, my gut feeling of "horse" will be slightly different than yours even though we accept the same scientific description of the animal. Subjective truth allows for error in perception as well, which may be the case if I perceive the horse while stricken with a case of yellow jaundice.

It should be noted that subjective truths also include aesthetic statements about the outer world. "That Andrew Wyeth painting is more beautiful than this one by Mildred Sands Kratz." Aesthetics can also range to even deeper feelings regarding the import of the natural world or the necessity of human love.

So, all these things considered, where does subjective knowing reside? First, it resides internal to an individual in the form of epistemic feelings. These feelings are embodied and physical in nature. They exist regarding a person's internal world as well as external things and aesthetic notions. And finally, where objective knowing finds its truth through verification by the outside world, subjective knowing finds its truth within as an internal confirmation.

Now, the third form in our topology is intersubjective knowing which is basically a social endeavor. It comes about among a group of people who agree. We have learned much about intersubjectivity from the work of the American theologian and process philosopher Joseph Bracken. In his book *Subjectivity, Objectivity, and Intersubjectivity: A New Paradigm for Religion and Science* he teaches that we as groups unconsciously co-create "a common field of activity" that strikes us as so common sense that it is experienced as objective reality, rather than dependent upon us and the other members of our group.

Indeed, the agreement concerning certain central beliefs of a group is often the cohesive force of the group itself. The agreement can become so axiomatic that the people see it as common sense, "We hold these truths to be self-evident, that all men are created equal, that they are endowed by their Creator with certain unalienable Rights, that among these are Life, Liberty and the pursuit of Happiness." It should also be kept in mind that intersubjective knowing can also be a shared illusion, as in case of the shackled people in Plato's cave who collectively believed in shadows.

The entire discussion of intersubjective knowing was introduced to the West in the early twentieth century by the German phenomenologist Edmund Husserl. In the *Cartesian Meditations* he calls intersubjective knowing an "empathetic experience" wherein we attribute willful actions to other human beings because they look like we do, behave much like we do, and give witness to similar traits as we do. Further, we hold that the individual view of the world of these others will also be similar to ours.

The views of the world of all the members of a group form into a common sense, usually unreflective, complex of meanings that make up the basis of a shared language and a shared experiential context that Husserl called a "homeworld" [*heimwelt*]. The homeworld is a consensus reality held together by interlocking conventions that make the people in the group aware of what is normal and serves as a source of epistemic confirmation for everyone.

Especially with this addition of Husserl's spatial notion of homeworld, we see now why the topological approach is so telling, so useful as a guide in understanding the three ways of knowing. So, we ask one more time: where does intersubjective truth reside? Where objective truth lies in the outer world of concrete things and subjective truth lies in the inner world of individual experience, intersubjective truth lies in the social world, in the relationships of human beings. Relationship confirms it. Intersubjective knowing becomes truth not through external verification, not through internal feeling, but via the relationship of people who hold the same truth in common.

Our explicit analysis of these three ways of knowing, these three epistemic pathways, has laid the foundation for another critique of the popularizer scientists. For, as we shall see, when it comes to their propositions regarding such important notions as the creation of the universe from Nothing and the visual experience of the world as being constituted of three spatial dimensions, they oftentimes conflate and confuse objective, subjective and intersubjective knowing. They especially, get lost in (un)thinking they are employing verified objective truth when in fact they are employing confirmed intersubjective truth.

Faith as Heuristic

To go one step farther in building Nihilology, it is incumbent upon us to fully illustrate just how intersubjective knowing is active within the realm of the religions of the world. In Sanskrit it is called *shraddha, iman* in Arabic, *emunah* in Hebrew, *fides* in Latin. These concepts overlap to such an extent throughout religious traditions, we will present our analysis using the single term English speakers are most familiar with: faith.

Now, faith is indeed only inconsequentially propositional in nature, in contrast to the case with objective knowing that can be expressed in the form, "I understand that..." Faith does make propositions such as, "The Four Noble Truths epitomize the Buddha's teaching," "God is trinitarian in nature," and "The *tetragrammaton* gives us the real name of the

God of Israel." Yet, these propositions are only a tertiary part, a prelude to the cardinal nature of faith that is not fully captured in propositions but believed intersubjectively by a group open to the truth of the proposition, what lies behind the Trinity, beyond the four noble truths, or within the four letters of God's name. This cardinal nature of faith is best expressed, not with "I understand that..." but with "I believe in..." The "in" also points to the fact that faith implies a commitment on the part of the believing group of people.

Father Daniel Berrigan, the famous and extremely influential Jesuit priest and disobedient was once infamously asked where faith lies. He replied, "Your faith is rarely where your head is at and rarely where your heart is at. Your faith is where your ass is at! Inside what commitments are you sitting?" Berrigan was, in his typical earthy manner, bringing home to the questioner that faith involves mentality and emotion and the will. What one thinks and what one feels is necessarily acted upon; without action it is not faith.

Objective knowing was seen to be a function of the mind (propositions). Subjective knowing was seen to be a function of the mind and the heart (epistemic emotions). Now we see that faith is an evolved form of intersubjective knowing that is a function of the mind, heart, and willful action. We like to say that faith is best epitomized as psycho-emotive gumption!

Further, let us take a look at the epistemic role of faith as we move along, step by step, toward our discussion of meta-conceptual knowing and ultimately to our culminating analysis of Nihilitive knowing. We will take this particular step in the shoes of an individual believer. With the culmination of this walk at Nihilitive knowing, we will ask one final time, "Where does Nihilitive knowing reside? Meanwhile, employing all that we learned in our analysis of the important role of heuristics to insights and discoveries in science, we will elucidate many parallels for faith as a heuristic in the realm of religions.

Back in the 1950s, Gordon Allport, the boundary breaking psychologist from Harvard, introduced us to the notion that people's religion matures in parallel with their psychological maturation. He held that the central feature of a mature religion is that it has a heuristic nature.

In earlier sections of our book, we presented the writings of Ivo Cohelho, Bernard Lonergan, and others who analyze how, in the attempt to understand Nothing, heuristic thinking (in the form of rules of thumb, pithy questions, and other problem solving tools) anticipates the understanding for which it strives. As we strive for deeper understanding of Nothing, there is an expectation in us of the general form of that understanding. We coined this expectation a "leaning into" that provides motivation to take on the search for Nothing.

Allport, in his book *The Individual and His Religion*, showed a unique discernment in focusing in on motivation as the key to religious maturation. He held that both children and adults could practice immature religion that is driven by impulses, fears, and magical wishes. what he calls "desires of the body." Children and adults can also practice mature religion that is "less of a servant, and more of a master," which urges a person toward a goal transcending self interest.

Whereas puerile faith is employed by believers, grown-up faith, rather, employs believers and drives them to discover new understanding beyond the bounds of current belief. As a result of this overturning of the employee/employer relationship, mature faith takes on the job of "master motive" for seeking new understanding.

Allport also discovered in his research that mature faith is open and unfixed. And, indeed it must be so, in order to work as a heuristic momentum toward, a leaning into, deeper understanding. "Feeling that we already have the one final and complete truth would not be heuristic and therefore would not be mature... the mature religious sentiment is ordinarily fashioned in the workshop of doubt."

This notion of faith as heuristic is not new to modern day psychological research. It is as old as the eleventh century and is well-articulated in the writings of—our now steadfast friend—Anselm of Canterbury. He frequently states that his contemplations on the nature of God are attempts to understand what he believes. "For I do not seek to understand in order to believe, but I believe in order to understand. For I believe even this: that unless I believe, I shall not understand."

Much like Gordon Allport, who reversed the employer/employee relationship of person and faith, so Anselm herein reverses the order of understanding and faith. In our modern-day culture dominated by the scientific worldview, our onus is to understand first and only then come to believe. This is often referred to in scientific theology as the rational basis of faith. Anselm held that it is the other way around; faith—made up of mind, emotion, and will—is the requisite condition of the person to bring forth understanding. The popularizers of science may want to pay attention to Anselm. They may be jeopardizing their understanding of the universe due to their staunch rejection of common faith!

We should also be reminded that, when Anselm was endeavoring toward contemplating God, he took for granted that contemplation was a knowing beyond concepts. As we quoted Anselm before, "Therefore, O Lord, not only are You that than which a greater cannot be thought, but You are also something greater than can be thought." Thus, faith points beyond itself and serves a heuristic function much akin to the apophatic theology we discussed earlier. Whereas apophatic theology uses words to negate the possibility of understanding God via mentality, faith subsumes the mind into an egalitarian relationship with emotion and the will. Together they form a ground wherein blooms meta-conceptual knowing of God.

Now, the notion that faith serves as an effective heuristic is not only voiced by religious leaders of the past centuries. Here are two examples from cognitive theologians of today. J.B. Gaskill, in his article "On Approach: Reflections on the G-d Discovery," advises believers to suspend any prior rejection of transcendent reality. "This threshold acceptance that there really is something much more than the mechanical level of explanation constitutes a heuristic mindset..." Gaskill employs the term "mechanical level" to refer to the realm of rational thought and advocates for accepting within oneself the possibility of the meta-mechanical.

Also, Ronald Kuipers, in his book *Critical Faith* harkens back to Gordon Allport's position that faith becomes the master motive in a believer's life. "[Faith] is like an obsession with a problem known to be insoluble, which yet follows, against reason, unswervingly, the heuristic command: Look at the unknown!"

Kuipers's poignant and inspiring simile recalls the discussion earlier about how images have served as heuristics in the history of scientific discovery. Our research revealed that many scientists found their way to insight using heuristics not involving words or even thoughts. They often used pictures, geometric diagrams, or three-dimensional models. We proposed that such images be assigned their own category that we term trans-heuristics, bridges that lead scientists to insight just over the cliff edge of rationality. We called on, among many others, the work of Arthur I. Miller who holds that images provide scientists with the escape velocity to break free of logic's gravity in order to reveal new insights.

We have now come to appreciate that faith functions as an effective trans-heuristic as well. Yet, whereas image is a heuristic of the mind's eye (Blind faith cannot function as a heuristic), faith is more than in the mind's eye. Faith is psycho-emotive gumption, an experiential ground from which one springs over the cliff of rational thought. As Anselm taught, it serves as a bridge to the meta-conceptual contemplation of God.

Yet faith, if we reflect on it for a long moment, cannot limit itself to meta-conceptual understanding of the realm of the divine. There is no boundary to the heuristic effect of faith. What sort of knowing will spring up, in what realm, cannot be predicted. Thus, we propose that it is reasonable and rational to commit oneself to a religious faith in order to lay the ground for scientific insight as well. Let the popularizer scientists take note. Their understanding of the universe has been enhanced via the employment of trans-heuristic images. Perhaps they should employ the common trans-heuristic of faith as well.

Meta-Conceptual

So, we have thus far elaborated with our topology the three sorts of knowing alive in the realm of concepts. Along the way we have offered a perspective on images and faith as trans-heuristics, bridges to wider sorts of knowing beyond concept. Now, we will take yet another step, a big one that thrusts us over the cliff edge of rationality.

A very precise language is required for this discussion in order to keep our points clear. This precision has been in force throughout our book thus far, but we need to make it explicit now so that our readers are cognizant of it, for, indeed, the structure of knowing is itself an aid in understanding the content of this discussion.

We employ the term "psyche" to refer to the realm consisting of the conscious and unconscious. The conscious is our daily world of open-eyed perception, joys, and problem solving. "Unconscious" has been approached and described in four different, yet related, ways. First, the "un" is not conscious, but what it actually is remains (acceptably for many science writers) indeterminate. Then there is the unconscious according to Freud and the school of psychoanalysis, portrayed as a stronghold of atavistic forces. Third, as the neuropsychologists say, the unconscious is attested to by right brain neurological excitation. Finally, according to cognitive psychologists, the unconscious is the seat of parallel processing of information in search of conscious problem solving. (We will discuss later the realm referred to by the term "super-psyche.")

Now, we refer to the realm of the unconscious as meta-conceptual in that rational thought has no life (*élan*) here and it also cannot access this realm. Nonetheless, certain prepared people can stay awake in the unconscious. Lucid dreamers, OBE adepts, and catharsists are three of the many documented examples.

Here we introduce a new and important term to our structure of knowing: being awake. This term allows us to differentiate between people who have meta-conceptual *élan* (life) without awareness and those who have meta-conceptual *élan* with awareness; the latter are awake. Our thanks to Buddhists for allowing us to borrow their extremely effective heuristic. The verbal root *budh* means "to awaken." From it we derive the name buddha, the "awakened one."

Let us, for one moment, turn our attention to a most famous insight in the history of science that illustrates the import and effectiveness of staying awake in the realm of the unconscious. In 1862, August Kekulé, the famed German chemist, saw an imaginary vision of a snake eating its own tail. This gave birth to his eureka! about the circular shape of the benzene molecule, the solution of which Kekulé had been working intently on for much of his young career. Almost thirty years after this root event, in a speech to the German Chemical Society, Kekulé finally described exactly how it came about.

One evening he was working on a textbook, but the writing was a trudge. His mind began to wander, so he went with it, turned to the fireplace, and shut his eyes. Atoms were "gamboling" before him, and he could see them very acutely in that he, by his own report, was disciplined in visions of this sort. He saw long lines of atoms twisting and wriggling about like snakes. "But look! What was that? One of the snakes had seized hold of its own tail, and the form whirled mockingly before my eyes. As if by a flash of lightning I awoke..."

We see here again, as we have proposed many times earlier, the essential role of images in scientific discovery. Furthermore, it is often told that Kekulé was sleeping and had a dream of the snake with its tail in its mouth. However, Kekulé himself clarified that his state was one of *traümerei,* not a simple sleeping dream (*traum*), rather what English speakers call reverie. (In the language of today's psychology, it is often termed creative imagination.)

The image of the self-circled snake is an ancient one. Usually given the Greek name *ouroboros,* it has been found in mythic texts and architecture of Egypt and India as well as in medieval alchemical works from Europe. It connotes endless rebirth or eternal return. Carl Jung, the prodigal son of Sigmund Freud, called it an archetype of the unconscious mind. (Archetypes are collectively inherited patterns or images that are universally present in individual psyches.) It was this unconscious *ouroboros* that Kekulé encountered in his reverie. And there is yet another piece of evidence from the Kekulé speech in 1890 that points to his being in *träumerei* (reverie) rather than *traum* (dream). Remember, he held that he was accustomed to this state of visions; that is, he had trained himself to make use of them for insights into his scientific investigations. So, Kekulé consciously made use of his unconscious for his work in chemistry!

Before we take our next step, to an even more rarefied realm of knowing, perhaps it is good to remain tethered to the earth when exploring scientific discovery. In describing his actions after awaking from his unconscious lightning bolt, Kekulé warns his listening audience of scholars, "... I spent the rest of the night in working out the rest of the hypothesis. Let us learn to dream, gentlemen, then perhaps we shall find the truth... but let us beware of publishing our dreams till they have been tested by waking understanding."

Thus, we see that August Kekulé was adept at staying awake in the meta-conceptual unconscious in order to manipulate images for the sake of scientific insight. Let us now revisit the teachings of the world's religions, as it is their contention that human beings can stay awake in a realm of knowing that lies wholly outside not only concept but outside the psyche itself. We term this realm of knowing the "super-psyche." In the realm of the super-psyche we will finally encounter Nothing.

Stay Awake!

Each religious tradition puts forth its own unique sort of training for its adherents to stay awake. It is certainly true that there are techniques in common across the traditions. We recognize, for instance, the kinship of prayer and meditation practices around the world and throughout time. Yet, the cultural and intellectual context out of which a religion is born, what we have called its birthing mud, lends a special and specifically effective strength to its ways of staying awake.

The Buddha preached that the human being is without a real self. He directed his disciples to concentrate on the constituents of the self-experience. They discovered behaviors, vital organs, memories, desires, feelings, a name... but no selfhood. The early Buddhists were all monks (*bhikkhu*) who were intense practitioners of a hyper-focused and, to modern western eyes, a perhaps picayune analysis of all aspects and functions of the mind and body in the vain attempt to find a self therein.

Here are the words of the Buddha from the canonical Samyutta Nikaya, extolling his monks to carry out mindfulness of their breath:

> How does a *bhikkhu* abide contemplating the body as a body?
> Here a *bhikkhu*, gone to the forest or to the root of a tree
> or to a room that is void, sits down; having folded his legs crosswise
> set his body erect, and established mindfulness in front of him
> just mindful he breathes in, mindful he breathes out
> So breathing in long, the *bhikkhu* understands "I breathe in long"
> or breathing out long, he understands "I breathe out long"
> breathing in short, he understands "I breathe in short"
> or breathing out short, he understands "I breathe out short"

He trains thus: "I shall breathe in experiencing the whole of breaths";

He trains thus: "I shall breathe out experiencing the whole of breaths"

We quote this passage at length to give our readers a taste and appreciation of the minute scrutiny of breath that this practice entails. Monks carried out this sort of analysis on walking, eating, relieving themselves, and thinking, all directed toward *abhisamaya*, "coming to completely" in a flash of awakening to the reality of no-self. The self is discovered to be Nothing (*shunya*).

The Zen Buddhists infused their minimalist cultural style into their *koan* wherein an aspiring student is confronted with an impossible conundrum and, through the status power of the master, is forced to respond. We recounted the story of Zen master Shuzan who embodied his famous *koan* (conundrum) in a lightning strike gesture. He held up his bamboo stick and challenged his disciples, "If you call this a bamboo stick, you oppose its reality. If you do not call it a bamboo stick, you ignore the fact. Now what do you wish to call it?" Students, in a heightened state of attention, must somehow express the reality of master's stick as unseparated from Nothing (*shunyata*), a feat that words cannot manage and must be accomplished with a mind gone silent with ineptitude. (This *koan* [conundrum] has often resulted in the famous whacking on the master's head... with his own stick at the hands of a disciple!)

Taoist sages employed austere, years-long retreats in remote hovels to practice quiescent meditation to the end of staying awake with a mind emptied of all waves. Before we quoted a verse from the root text of the tradition that pointed to the empty hub of a wheel as the source of the wheel's effectiveness. A commentary then reflected, "Likewise, it is because the minds of sages are nothing [*wu*] that people turn to them for help." Zhuangzi, the fifth century BCE philosopher expanded by noting when water is completely still it reflects perfectly the face of they who gaze into its mirror-like surface. "Such is the clearness of still water, and how much greater is that of the human Spirit! The still mind of the sage is the mirror of heaven and earth..."

"Heaven and earth" is a merism akin to those in the biblical book of Genesis. The phrase connotes the whole universe. According to the tradition of Taoism, the mind that is Nothing [*wu*] knows everything and it does this by reflecting everything without any waves of its own.

Now, several hundred years after the first Taoists and half a continent away, out of a quite alien birthing mud, arose the Hindu tradition of Samkhya. Unlike the Taoists, Samkhyans did not advocate a renunciate, staunch quietism to stay awake in the realm of super-psyche. Rather, their meditation was manly an open-eyed observation of the world amidst all its activity.

Recall, Samkhya teaches that the basic ignorance of human beings is that we falsely identify our self with the created things of the world. Further, they show great faith in stating that the world is only made manifest for us to observe it. The Samkhya way of staying awake is to observe intently all aspects of manifest life with the active internal question "Am I that?" With the final "No" practitioners free themselves from ignorant false identification with the world, free from experiencing possessions, body, emotions, and all mentality as "me." Free from ignorance, humans again assume their real identity as consciousness only. True selfhood, according Samkhya, **is** the super-psyche.

The Buddhists of the Mahayana school created their own form of *via negativa* practice. Recall the poetic words of Nagarjuna, "The victorious ones have said that nothing [*shunyata*] is the relinquishing of all views. For whomever nothing is a view, that one will accomplish *nil.*" The

Mahayanists evolved into great debaters, entering the rule-guided fray of words as a "destroyer," attacking all intellectuality regarding reality. The attack reduced the opponent's concepts to absurdity, disallowing them their hypostatization, i.e., making words into existents. With the loss of real connection between concepts and things, in the intense psychological heat of debate, the opponent is introduced to *nirvana,* the sought-after condition wherein concepts are blown out as by a cooling wind *(nir-va)* and Nothing is known.

In the West, Christians brought forth their unique practice of the *via negativa* as a way to stay awake in the super-psyche. We visited it earlier in the writings of the apophatics, theologians of negation such as John of Damascus. "These words about God do not show what He is, but rather, what He is not." Or John Scotus Eriugena, "We may learn from creatures *that* God is but not *what* God is." God, according to Eriugena is at once Super-god (*super deus*) and Nothing (*nihil*), spoken thus to negate the power of theology to capture the divine nature. Yet the apophatics continued to speak, because speaking about the Nothing of God was held to be an effective way of going beyond the very words uttered.

It must be noted that Eriugena held that certain rare humans indeed had super-psychic capacity, when aided by divine manifestation (*theophania*), to enter into a condition of deification (*theosis*) wherein the person is utterly in identity with God as Nothing. This identity is not a trope for Eriugena, as he took the statement of the fourth century Augustine of Hippo quite literally. "God was made man so that man could be made God." Let us again quote from a sermon of Meister Eckhart, an heir of Eriugena, to cement soundly this teaching of *theosis*. "The eye with which I see God is the same with which God sees me. My eye and God's eye is one eye, and one sight, and one knowledge, and one love."

Further in order for all of us to gain an appreciation of just how apophatic speaking can possibly function as a way, we explored the research of neurotheologians Andrew Newberg and Mark Robert Waldman, neuropsychologist Steven Lehar, and others. Their science shows that words also have effect completely outside of meaning, exhibiting the inherent property of altering the makeup of our brains, restructuring the thalamus, and enhancing how we perceive reality. Further, when one listens empathetically to another, both people develop a neural resonance with each other; each mirrors the neurological *élan* (life) of the other. Thus, neurological science helps us understand that, when an apophatic theologian resides in the divine (Nothing) and speaks to listeners about the divine (nothingness), the minds of the listeners begin to reflect the mind of the speaker. Thereby words function to bring listeners into the realm of Nothing as well.

And finally, we found much of the *via negativa* expressed in the teachings of Judaism. The more mystical strains of the tradition, especially witnessed in the writings of Kabbalah, were unashamed to advocate for specific techniques, exact ways to prepare adherents to stay awake in the super-psyche.

Yahweh, in his most rarefied realm, is called *Ein Sof* (limitless) or *Ayin* (Nothing). Recall the words of David ben Judah he-Hasid as to knowing this limitless Nothing, "...therefore forgetting pertains to the comprehension of this realm." And Issachar Ber of Zlotchov, "... through realizing that there is no place empty of [Yahweh]... one comes to the state of *ayin*... and is contained... in the creator." Now, how does Kabbalah describe the way of forgetting and becoming Nothing? By contemplation of the *sefirot.*

These ten creative qualities give witness to Yahweh's manifestation into the world. They are often portrayed in a symbolic configuration that the tradition calls the Tree of life. Practitioners of Kaballah envision the tree and contemplate the ten qualities. Beginning at the base of the trunk with *shekhinah* (Yahweh's presence in the world) they climb up through mercy and understanding to the topmost leaves of *keter* (divine crown). *Keter* is often said to be the highest *sefirah* comprehensible by human mentality, yet it is not the practitioner's final goal. We look to Daniel Matt once again, this time in his book *The Essential Kabbalah: The Heart of Jewish Mysticism* where he tells us there is a higher, deeper realm beyond this paced step approach. "At the ultimate stage, the kabbalist no longer differentiates one thing from another. Conceptual thought, with all its distinctions and connections, dissolves."

The practitioner no longer analyzes the ten *sefirot*, and no longer differentiates their self from the self of Yaheh; both are *ayin*. And their shared Nothing is known in the realm of the super-psyche.

Now, there are resounding echoes of the Jewish Kaballah heard among the mystical schools of Islam when teaching ways of staying awake in the super-psyche. Recall our discussion before of how Al-Ghazali called life on earth "an imprisonment in clay" where sensual experience puts a veil between humans and Allah. He taught Muslims to recollect their immanent death twenty times a day so that, when the pot was actually broken, they could have "the delight of vision" of Allah, unite with him as "friend to friend," and assume their "union with him." With a closer listen to Al Ghazali, however, it becomes quite clear that these teachings were spoken for the ears of common Muslims who perform *salah* five times a day and go the mosque at noon on Friday.

The other ears listening to Al Ghazali were his close brethren in the order of Sufis. Historically, Sufism has witnessed the Prophet's teachings of unmediated encounter with Allah. By the late eleventh century, when Al Ghazali came into religious maturity, its influence was spread throughout the Middle Eastern world. He called the Sufis true *arfin* (knowers) of Allah. Compelled by his own transformative experience as a knower of Allah, he spent ten years in prayerful seclusion before emerging as a leader of the Sufi order in Damascus and later in Baghdad.

According to Claud Field in his preface to the translation of *The Alchemy of Happiness,* Al Ghazali in his own day was known for teaching "one doctrine for the multitude and one for himself and his intimate friends." He revealed a perspective on Muslim life to fellow Sufis that, while akin to his teachings to mosque-going Muslims, spoke openly of a true union with Allah while alive as a human on earth. For instance, Al Ghazali taught that "Some Sufis have had the unseen world of heaven and hell revealed to them when in a state of death-like trance. On their recovering consciousness their faces betray the nature of the revelations they have had by marks of joy or terror."

Again the use of a merism as we have seen many times in texts of religions from around the world. "World of heaven and hell" refers to the entire realm entered at a person's death. "Death-like trance[s]" are the meditative states Sufis engender. (Whether one returns to daily consciousness with a look of joy or terror on one's face depends upon the degree of skill fostered in the practice of d*hikr* [remembrance], as we shall discuss soon.) So, the very "delight of vision" of Allah that Al Ghazali taught to mosque-going Muslims as possible upon physical death, we now hear taught to Sufis as attainable while very much alive!

Now, Al Ghazali also calls this delight of vision of Allah a "uniting of friend to friend" and a "union with Him." For our discussion of ways of staying awake in the super-psyche we must elaborate more specifically the nature of this union and the means of attaining it. Al Ghazali teaches for the ears of mosque-going Muslims as distinct from Sufis in this regard as well.

He recalls the words of the Prophet "He who knows himself, knows God." That is, by contemplation of one's own being and attributes one arrives at knowledge of God, and there must exist a special means of doing so. "As a matter of fact, there are two methods of arriving at this knowledge, but one is so abstruse that it is not adapted to ordinary intelligences, and therefore is better left unexplained."

Thus, Al Ghazali teaches two ways of knowing Allah. He recommends for mosque-going Muslims to reflect on the mercy of Allah as witnessed in the created world. The other way he considered "abstruse... not adapted to ordinary intelligences." Just as was the case with the proclamation of true and ultimate unity with Allah, Al Ghazali held that the practice of the other way to stay awake would be dangerous and misused by mosque-going Muslims. However, he was a practitioner of the second way himself and it is well-documented in many Islamic texts and commentaries, especially those of the Sufi orders. This second way is called *Dhikr.*

D*hikr* means "reminding oneself" or "remembrance." It refers to a means of staying awake through repetition, either spoken externally as prayer or internally toned in silence. The content of *dhikr* is the names of God or short declarations of his nature. For example, the name Allah (the god, the Nothing), the urmotto of Islam *allahu akbar* (God is great), or the expression of gratitude *alhamdulillah* (All praise is due to God) may serve as the basis of remembering.

Let us listen to the Persian poet and Sufi teacher Jalaluddin Rumi, for he was a foremost teacher of *dhikr* (remembrance) as a way to stay awake in the super-psyche. Before entering into the way of *dhikr,* he warns that the recitation of a divine name will have no benefit if the practitioner continues in their daily thinking "of cows and donkeys" while repeating the divine name with only the tongue engaged. (Note again the merism "cows and donkeys," which points to the whole world of daily thinking and activity.) He then delivers his poetic vision of how *dhikr* [*zikr*] really finds its mark:

> A naked man jumps in the river, hornets swarming
> above him, The water is the *zikr*, remembering,
> There is no reality but God. There is only God.
>
> The hornets are his sexual remembering, this woman,
> that woman. Or if a woman, this man, that.
> The head comes up. They sting.
>
> Breathe water. Become river head to foot.
> Hornets leave you alone then. Even if you're far
> from the river, they pay no attention.
>
> No one looks for stars when the sun's out.
> A person blended into God does not disappear. He, or she,
> is just completely soaked in God's qualities.

The practitioner of *dhikr* allows the names and natures of Allah to completely wash over their selves, the ideas and fantasies of life on the shore. "The hornets of sexual remembering" no longer intrude. The names and natures of Allah "completely soak" the practitioner until there is only the river. (Remember the Jewish technique of contemplating the *sephirot*.) The practitioner doesn't cease, yet they no longer have separate existence. Nobody, including them, will see their star because the *dhikr* has come out as the sun.

So now that we have elucidated these many ways of staying awake as witnessed in the religious traditions of the world, what have we learned?

First, religions do not leave this staying awake to chance, fate, grace, or the divinities. They teach specific methods to accomplish awakeness. Second, the traditions do not treat the realm of the super-psyche as a mysterious, shadowy place. No light can be brought to this realm by mentality (that has the power only to lead us to the entrance gate but whose strength fizzles at the threshold). However, this in no way implies that the super-psyche is beyond the strength of the prepared practitioner. Indeed, the hallmark of religious traditions may be their comfort and facility with the fact of being awake in the super-psyche. And finally, we have learned that the super-psyche is the realm of the divine as Nothing.

Confluence of Writings on Nothing

Let us now visit one more time the religions of the world where we discover a confluence of teachings, both among religious traditions and between religions and science, as to the exact nature of knowing Nothing. From these confluences we will build more heuristic features of Nihilology.

In an earlier discussion, we introduced what we termed the rarefying principle. The principle states that the nature of created things is more and more revealed by the shedding of excess matter. It served as a handy heuristic in our understanding how scientists have gone about answering the question of the makeup of the manifest world.

The history of the scientific quest to reveal the essence of the atom highlighted the principle. Around 1900 Thomson proposed that an atom consists of negatively charged corpuscles floating in a positively charged ooze. In 1910 Rutherford swapped the heavy term "corpuscle" for the energetic "electron" and declared there was no ooze at all;

the electrons flew through empty space around a massive positively charged nucleus. In the 1960s it was discovered that protons and neutrons of the nucleus are not elementary particles at all. Rather they are composed of quarks that are surrounded by a vast space within the larger particles. Then, in the 1970s, it was proposed that particles in the atom were not three-dimensional points in space but more akin to vibrating strings of a single dimension. Matter had been reduced to a sort of elemental rhythm. Thus, the principle: as a created thing becomes more and more understood, it turns out to be less and less dense; it sheds matter, becomes rarer.

This rarefying principle from science is also a handy heuristic in understanding how religions of the world have gone about answering questions regarding the nature of the human being and the nature of the divine. It will guide us as we investigate the evolution of teachings expressed in the diverse traditions: from faith of the every-Sunday congregation and mosque-goers with its focus on mythic truth and orthopraxy, up to mature believers motivated by the necessity to confront the unknown mysteries, to the sophisticated *apophasis* of professional theologians and reflective mystics. In this evolution, we indeed find the shedding of matter from the essence of the human and the divine. Yet, this shedding is much more profound than the simply material.

Regarding the essence of the human being, we will hear religious teachings steadily and systematically stripping us of body, possessions, thoughts, aspirations, individuality, and selfhood. The divine is also not spared this radical paring down: no body, no voice, no personality, no form, no being. Over the cliff-edge of this evolution the rarefying principle will find its denouement. There, religions teach us that, in their most rarefied, ultimate form, human nature and the divine nature are the same: Nothing.

Now, the Buddha was born into a Hindu religious milieu that put forth as its basic teaching that human beings were an essential soul (*atma*) in transmigration (*samsara*) from life to life, reaping the rewards and debts of behaviors and attitudes. At his great awakening he saw that all his suffering as a human was rooted in the false experience of living through an individual soul. In shedding his soul, he was left in the state of nirvana, the bare condition wherein all thinking, false experience, and thirsting after rebirth are blown away as with a cooling wind. This is the most rarefied and ultimate condition of the soul as Nothing (*shunya*).

The goal of Buddhism throughout its 2,500-year history has been to enter into the condition of nirvana. Certainly, there have always been Buddhists who light incense in the temples devoted to the many divine beings who reign over the quadrants of heaven, imploring them to aid the worshipers to join them there for not-quite-eternal eons of joy. For these incense-burning Buddhists, nirvana remains a far-off, unformed, untouchable place. However, for early monks (*bhikkhu*) sitting cross-legged in the forests with the Buddha, as well as for teachers and contemplatives ever since, the divine goal of their Buddhism has never included, nor been aided by, divine beings. Their attainable ultimate nirvana is indeed an impersonal state.

As Buddhism evolved, it boldly ballooned the no-self notion out and out until it incorporated all existents, ultimately stating that all life is without essential, independent reality. All life is *shunya* (Nothing). Let us once more quote verses of Nagarjuna, the great debater of later Buddhism:

> There is not the slightest difference
> Between cyclic existence [*samsara*] and nirvana
> There is not the slightest difference
> Between nirvana and cyclic existence

Whatever is the essence of the *Tathagata*
That is the essence of the world
The *Tathagata* has no essence
The world is without essence

Whereas Hindus had taught that the goal of life was to free oneself from the cycles of birth and death, Nagarjuna gives voice here in the first verse to the teachings of later Buddhism (*Mahayana*) that the goal is to see, in the most rarefied form of reality, no difference between the worlds; both the ignorant realm of moving life to life and the realm of "blown away" are Nothing (*shunya*).

In the second verse, Nagarjuna also voices an evolution from early Buddhism. *Tathagata* was a third-person reference that Buddha gave to himself, presumably to avoid the use of "I." It carries the meaning "one who is thus gone", i.e., unto the state of nirvana. Over a period of a few hundred years, the term ballooned in meaning to refer to the *élan* of ultimate reality, "goings-on of suchness." Again, in their most rarefied form, the nature of the daily world and the nature of the ultimate world are "without essence." They are Nothing (*shunya*).

Thus, Buddhism proposes that the individual human being is constituted of no essence, and the divine reality is constituted of no essence. They, in their most stripped-down nature, are both Nothing.

Further, as Buddhism made its migration across China to Korea and across the strait to Japan and there evolved into the sects called Zen, the tradition held fast to these teachings that individual self and world are empty of essence; they are *shunya*. And, as we discussed before, the masters of Zen, taking advantage of their authority, presented their disciples with *koan,* conundrums that frustrate the mind into ineptitude and demand a super-psychic response witnesses the disciple's opening to *shunyata*.

The most famous collection of *koan* (conundrum) is named *Mumonkan*, in one sense after the collector, a Chinese monk called Mumon (Chinese: *Wumen*). The collection came to be in the thirteenth century, the same era that saw Zen rise to maturity in Japan. *Mumonkan* literally means "the gateless barrier." Yet, in Zen, *mu* is equivalent to *shunyata* of earlier Buddhism, i. e. Nothing. Thus, *Mumonkan* can also mean "gateway to Nothing." The very first *koan* (conundrum) is famous throughout the world today, and no Zen practitioner receives the imprimatur of their master without solving this one: A monk asked Joshu. "Has a dog the Buddhanature?" Joshu answered, "Mu."

Now Joshu has indeed presented this monk a barrier without a gate, for it was a long-standing tenet of Zen that the Buddhanature permeates all things. ("Buddhanature" translates the Sanskrit *tathagatagarbha*, "inner suchness" or "womb of reality.") Our monk is, seemingly, asking the great ninth century Chinese master Joshu (Chinese: *Zhaozhou*) to simply confirm the basics. Yet, Joshu retorts in the most precise way to halt the monk's questing momentum: Mu!

Mu in Japanese is equivalent to *wu* in Chinese. We have seen this term in our discussion of Taoism. Both mean "not," "empty," or "Nothing." Mumon then warns us not to hold to a nihilistic concept of vacancy or a relative concept like "has" or "has not." "It will be just as if you swallow a red-hot iron ball, which you cannot spit out even if you try... Internal and external will be spontaneously united... then all of a sudden an explosive conversion will occur..."

Mumon is warning all those who take on this root *koan* that the solution lies beyond all concept. He later emphasizes this warning in the typical confrontational style of Zen, "When you meet Buddha, kill him!" Contemplating the *koan* must be a vise of heat; no images or psychic activity can survive. In the realm of the super-psyche, the self and the world become non-differentiated, and an "explosive conversion" will occur.

A more literal translation than "explosive conversion" would be "*mu* will break open." As Mummon says, "In order to master Zen, you must pass through the barrier of the patriarchs... [this] single word *mu*." Remember *mu* also means "Nothing." So, "explosive conversion" also carries the meaning "Nothing will break open."

Now, it has been a unique convention among Japanese Zen masters to deliver a so-called death poem to their disciples at the moment of their passing. It is supposed to be the pithy sum of the master's understanding. Here is Mumon's death poem:

The void [*mu*] is unborn
The void does not perish
If you know the void
you and the void are not different

So, once again we see the confluence of these notions: in their most rarefied nature, the Nothing of self and the Nothing of the divine reality are seen to be one.

Earlier, we noted the echoes in Taoism of the Buddhist teaching that Nothing and creation are eternally present, that Nothing is not lost at the creation of the universe. And now we also hear these two traditions resonate in their teachings regarding the self and the most rarefied divine reality.

Remember, the Taoists employed the evocative image of the empty hub being the genius of the wheel and the empty (*wu*) mind of the sage being the genius of the human being. To attain this *wu* mind, students practice an extreme quiescence called *zuowang*, literally "sitting-forgetting." Livia Kohn, perhaps the most prolific writer in the West on the practices of Taoism, tells us that sitting-forgetting is a meditative absorption wherein sensation and reflection are quietened. This absorption rest is the key requisite for merging with tao.

The classic description of sitting-forgetting is found in the writings of the fourth century BCE Zhuangzi, the sage we quoted before who is accepted as the greatest voice of Taoism after the "Old Master Lao" himself. In this vignette, a student named Yan Hui is talking with his master Zhongni (Confucius).

Yan Hui said, "I am making progress." Zhongni replied
"What do you mean?"
"I have ceased to think of benevolence and righteousness," was the reply.
"Very well; but that is not enough."

Another day, Hui Hui again saw Zhongni, and said, "I am making progress."
"What do you mean?"
"I have lost all thought of ceremonies and music."
"Very well, but that is not enough."

A third day, Hui again saw (the Master), and said, "I am making progress."
"What do you mean?"
"I sit and forget everything." Zhongni changed countenance, and said
"What do you mean by saying that you sit and forget (everything)?"

Yan Hui replied, "My connexion with the body and its parts is dissolved
My perceptive organs are discarded
Thus leaving my material form, and bidding farewell to my knowledge
I am become one with the Great Pervader."

"Great Pervader" more literally is translated as "big tao." Tao is the way that permeates the universe. It is *wu* in movement. Thus, we often hear the phrase *wu-wei* in a sort of *apophasis* regarding tao. Literally this means "Nothing doing," i.e., the *élan* of Nothing.

Evan Morgan, in his now famous early work *Tao, The Great Luminant: Essays from Huai Nan Tzu,* gives us a fertile description. W*u* and *tao* are equivalents. With this sameness in mind, it can be said that *wu wei,* no action, is *tao wei,* action by the spirit way. So, the goal of Taoist sitting-forgetting is to become non-different from *wu*, non-different from the tao. The empty mind is identical with the great empty. Nothing greeting Nothing.

Now, let us again return to two more religions of the East, both sprouting out of the Hindu tradition, to further illustrate this rarefying principle that reveals the sameness of the self and the ultimate divine.

We discussed before the *apophasis* of *Advaita Vedanta* as witnessed in their mantra motto: n*eti neti*, "not this not this," that strips the human being of personhood attributes and *Brahman* of divine attributes to clear the path to non-differentiation, hence *Advaita*, or "non-dual."

This stripping of the human being is poetically pointed out in *The Crest Jewel of Wisdom*, the most important work of Shankara, the original spokesman of non-dual *Vedanta*:

Formed of the substances they call marrow, bone, fat, flesh, blood, skin and over-skin
Fitted with greater and lesser limbs, feet, breast, trunk, arms, back, head
This is called the physical vesture by the wise
The vesture whose authority as "I" and "my" is declared to be a delusion

The term "vesture" in this stanza translates the Sanskrit *kosha*, literally "sheath." *Crest Jewel* tells us that humans, fooled by the illusion dance of *Maya*, experience these sheaths as aspects of real selfhood. Yet, when freed from this illusive experience, the sheaths are seen as merely scabbards over their true nature as Self (*atma*).

The sheath rhapsodized above is physical, made up of fat, flesh, feet, and lesser limbs. Three others are made up of the rational activity, sensual experience, and vital breath The fifth is called *anandamaya,* "composed of bliss." This sheath is ghostlike and is only experienced by practitioners who can stay awake in the condition normally called deep sleep, very rare, yet a false scabbard covering *atma* nonetheless. And, according to Shankara, all five sheaths are "declared to be a delusion." Free of delusion, *atma* is freed from physicality, breath, sensation, thinking, and bliss; it is stripped down to Nothing.

Now, to refresh our memories, how does *Advaita* characterize the divine nature, the *Nirguna Brahman*? "*Nirguna*" means "without qualities." The Sansksrit name *Brahman* finds its roots in "expansion" and implies that *Brahman* is infinite expansion, infinity, not a personal deity at all, rather an impersonal divine.

Let us once again quote David Paul Boaz to elaborate on the infinite divine. "It is One, without a second, without limit, empty of all predicates, attributes and qualities, beyond concept and belief, or any subject-object dualism whatsoever... *Nirguna Brahman* [has] no ultimate, permanent essence, no essential self-nature." The divine nature, when made its most rare by stripping it of any attribute whatsoever, including the attribute of self-existence, is found to be Nothing.

Now, the name *Vedanta* means the "culmination of knowledge" and refers to the *Upanishads,* the culminating books of the sacred scriptures that Hindus refer to as The Vedas. Within the *Upanishads*, we hear what the tradition has come to call *mahavakya* or "great utterances." They sound to Hindus like "In the beginning, God created the heavens and the earth." Or "I and my Father are one." Or "Allahu Akbar." They are essential statements of the essence of the tradition.

Three of the four *mahavakya* declare the relationship of *atma* to *Nirguna Brahman*; this dominance of the great utterances witnesses the importance of this particular tenet in the culminating view of The Vedas. The first utterance is "*Ayam atma brahma*" or "This very self is the ultimate divine." The speaker here is a theologian who is concluding an elucidation of the nature of the self. The second utterance is "*Tat tvam asi*" or "You are, indeed, that." In this instance the speaker is a master speaking personally to a disciple as a capstone to their teachings on divine nature as that (*tat*). The third utterance is "*Aham brahmasmi*" or "I am the ultimate divine." This is the eureka! cry of a disciple as they awaken.

Thus, these *mahavakya* speak, in the voices of theologian, master, and disciple, the very same *Advaita* teaching; the self of the human being (*atma*) and the divine (*Nirguna Brahman*) are non-dual. Again, stripped of the five sheaths and all movements of the psyche, *atma* is indeed Nothing. And negated of any attribute whatsoever, *Nirguna Brahman* is left as Nothing. According to *Advaita Vedanta*, the most rare human being and the most rare divine being are not different; they lose differentiation in Nothing.

Now, the second school sprouting up within the Hindu tradition that we need to revisit is *Samkhya*. As it reveals its teachings on the nature of the human being and the divine, we hear a truly unique vision of the essence of consciousness.

Recall that the basic flaw in the universe, according to *Samkhya*, is that a pervasive ignorance dupes human beings into conflating *Prakriti,* manifest life, and *Purusha,* pure consciousness. The duped human experiences their life in a false manner, "This I is conscious." "I have emotions." "I exert willed actions."

In order to see through this duped life, one must differentiate *Purusha* from *Prakriti*, both of which are real, but only *Purusha* is the real human. The way to effect this differentiation is to practice what we have termed an open-eyed observation of the world in all its activity, what *Samkhya* calls *tattvabhasya* or practice on the essences (of manifestation). Verse sixty-four of the *Samkhyakarika* states, "As a result of practice on the essences, wisdom arises of the form, "I am not, not mine, the 'I' is not..."

For the seeker who sees through ignorance, the experiences they thought constituted their individuality turn out to be false. The things they thought belonged to them—body, emotions, will, houses, water buffalo, memories, insights—are not them. The very basis of egoity, what Samkhya calls the "I-maker" is actually not them at all. The seeker is other than the processes that take place around or within the body. They are not the mentalized movements of the psyche, in fact not part of existence at all.

As a result of a scrupulous observation of the world, the seeker finds the world to be totally other. The seeker is not the world in any form. The seeker is that which remains after all existents are stripped from them as self. The seeker has no positive existence; they are Nothing. The seeker is *Purusha*.

Purusha is literally translated as "human" and is said to be *cetana* alone. Now, *cetana* is normally translated in philosophical analyses as "consciousness," and we have conformed to this convention throughout our writing. Yet we must see that *Samkhya* teaches that *Prakriti,* the existents of the world that are totally **not** *Purusha*, possesses all the characteristics of what we normally assume constitutes consciousness. She is extremely insightful, able to make the subtle distinction between herself and *Purusha.* She is compassionate and self-giving. She is capable of direction and purpose. She takes charge of coordinating the karma mechanics of the entire universe.

Thus *Purusha* (human) finds essence in *cetana* (consciousness), yet *cetana* is not any form of mentality, insight, experience, or emotive empathy, **not** any existent! So, what is *cetana*? Verses sixty-five through sixty-six of the *Samkhyakarika* answer our question in describing the eureka! moment of the seeker:

> With this [vision] *Purusha*, like a spectator seated in his own seat, sees *Prakriti*....

> The uninvolved one [*Purusha*] says: "She has been seen by me"
> The other [*Prakriti*] says: "I have been seen."

Cetana is the bare fact of witnessing. Remember, according to *Samkhya's* anthropomorphic mythology, *Purusha* watches *Prakriti* dance in order to free himself from false identification with her as the world. *Purusha* himself is actually "uninvolved" in the dance. The verbal root of *cetana* is *cit,* ""to be awake." *Purusha* is the essence of *cit,* the state of awakeness. The nature of I, stripped of the world, is left with being the rarefied divine as Nothing... and Nothing is awake!

So, let us return one final time to the religions of the West to discover how our rarefying principle is active in traditions closer to home.

We discussed earlier the writings of the great ninth century Christian theologian Eriugena who stated baldly, "God Himself does not know what He is because he is not anything. Literally, God is not, because He transcends being." Eriugena also held that God is beyond the human capacity to sense or confirm mentally, so actually God is "nothing, due to the excellence" of His nature.

Nonetheless, even though God, in his most rarefied nature, is Nothing, Eriugena averred, with no sense of metaphoric speech, that some rare human beings can actually enter into the super-psychic realm of deification (*theosis*) where the human is utterly identified with God as Nothing. Eriugena holds that human beings must not to believe that God and themselves are distinct from one another but one unity. Human beings, by subsisting, are in God. And God, by manifesting himself in a mysterious and ineffable manner, creates himself in the human.

We quoted before the theology of Meister Eckhart, the thirteenth-century heir of Eriugena. Now, let us delve a bit deeper to make more subtle this Christian teaching on unity and to point out a few echoes of other traditions in this regard. In his sermon titled "Sanctification," Eckhart gives us a sense of how unity of human being with God may arise. "[T]he spirit remains as immovable and unaffected by all impact of love or hate, joy or sorrow, honour or shame, as a huge mountain is unstirred by a gentle breeze."

Please note that we have heard this exact teaching for the need of utter rock-like psychic immovability in the Taoist religion. Recall this pronouncement from Zhuangzi, "The still mind of the sage is the mirror of heaven and earth... all things are not able to disturb their minds..."

Eckhart continues with his preaching to say if the human heart is vacant of all earthly concerns it is ready for God. "If I wish to write on a white tablet, whatever else is written on the tablet, however noble its purport, is a hindrance to me... I must wipe the tablet clean of everything, and the tablet is most suitable for my purpose when it is blank."

Note again the echo here of the teachings about the process of human emptying (*kenosis*) that we heard in the Gospels, the writings of Carl Jung, and the poesy of St. John of the Cross. "It is meet then that the soul first be brought into emptiness and poverty of spirit and purged..."

And finally, Eckhart concludes his sermon with the goal of this rock-like psyche that has been emptied. "When the soul gets to this point, it loses its own distinctiveness, and vanishes in God as the crimson of sunrise disappears in the sun. To this goal only pure sanctification can arrive."

In another of Eckhart's sermons, he describes this same goal, this time employing the trope of the light of the soul unifying with the light of God, "When the soul enters the light that is pure, she falls so far from her own created somethingness [*das ist*] into her nothingness [*das nicht*] that in this nothingness she can no longer return to that created somethingness by her own power"

So, from these voices of the Christian tradition, we hear the rarefying principle intoned. God's nature, when seen in his most excellent condition, is so rare as to be Nothing. When the human being is emptied of all self characteristics, when the psyche is immovable and the heart is devoid of personal affections, the human is prepared for God to enter. By employing the most rare of physical images, we are told that the light of God enters the human being and they become unified as one light. The human "falls far from *das ist*" (that which is) and "falls into *das nicht*" (that which is not); the human as Nothing becomes one with God as Nothing.

Now, we presented earlier the teachings of mystical Judaism on the most rare form of God as Nothing (*ayin*). We quoted the words of theologians from across Jewish history: David ben Abraham ha-Lavan, Joseph Gikatilla, and Issachar Ber of Zlotchov. Each in their own language attests to the same ultimate fact: the most rarefied nature of the divine being is beyond concept, beyond characteristic, beyond all existents. God is *ayin.*

The process of self-emptying required to meet God as *ayin* is referred to in mystical Judaism as *bitul*, or nullification. It refers to the evolutionary process of stripping oneself of some-thing-ness (*bitul hayesh*) as well as the ultimate goal of super-psychic no-thing-ness (*bitul b'metziut*). This is how Rabbi Yitzchak Ginsburgh, a renowned contemporary integrator of mainstream Rabbinic Judaism and mystical Hassidism characterizes *bitul* to mean that the more one enters into wisdom the less they will experience their ego. "As the intensity of the feeling of one's own ego diminishes it is replaced by an acute awareness of the omnipresence of the Almighty."

One final input will help us appreciate *bitul* more subtly, this time from a more traditional source, the Rabbi Yitzhaq Yehiel Safrin who was an early nineteenth century leading exponent of the Baal Shem Tov, the founder a century before of Hasidism. The Tzaddiqim he refers to are "righteous ones" who may become great teachers of the tradition if they succeed at becoming low, "to become nullified... to be literally,

literally like a small creeping creature... literally as nothing... as dust and ashes."

This debasing to naught of selfhood for the sake of divine unity is also witnessed in Islam. Recall that Islam teaches the means of staying awake called *dhikr*, a contemplative repetition of the attributes of Allah. The goal of this contemplation is *fanaa*, the Arabic term meaning annihilation. For canonical imprimatur, teachers point to the famous saying of the Prophet, "Die before you die." *Fanaa* connotes the dissolution of individual identity in union with Allah.

Another Arabic term central to our discussion here is *tawhid*, meaning oneness. It is the absolute ground of Islam regarding Allah, God is one. Islamic Sufis expand this ground to teach that there is an intrinsic unity or *tawhid* between Allah and human beings. Remember the apophatic statements of the great Al Ghazali: Allah is beyond all concept, all psyche. Also, humans, in their most stripped down, most rarefied nature, are the reflection of Allah and Allah is the reflection of human beings. When naughted via the way of *dhikr*, humans, as Nothing, reflect the Nothing of Allah.

Now, Abu Yazid al-Bistami was an eighth-century Persian Sufi who is credited with being the first in the history of Islam to teach specifically of *fanaa* (annihilation). This story survives in the traditional literature:

Abu Yazid al-Bistami approached the Divine Presence and knocked on the gate
He was asked, "Who is there?" "I have come, Oh my Lord", replied Abu Yazid.
He was told: "There isn't any place here for two. Leave your self behind and come."

When Abu Yazid once again approached the Divine Presence and was asked who it was,
he said: "You, oh Lord."

Abu Yazid was halted at the gate to the divine as long as his *fanaa* was still in progress. Only when Abu Yazid died before he died was he allowed to enter Allah's house. This moral of the tale is taught in a more theological language by the foremost Sufi of all history, the late twelfth century Spanish teacher Ibn Arabi who we shed light on earlier.

In his treatise titled "The Book of Annihilation in Contemplation" (*Kitâb al-fânâ' fi-l mushâhadah*) he concludes, "... if you are not, you see Him... His vision does not occur except through your extinction from yourself."

So, we have witnessed a most clear confluence of notions from across the religious traditions of the East and West, each embodying the rarefying principle. The traditions of the East, at base, portray an individual self that is unreal from the start. Buddhism teaches that pervasive ignorance creates the experience of self, and overcoming ignorance reveals the self and the divine to be *shunya.* Zen offers *koan* to blunt all psychic strategies toward an explosive revelation of *mu.* Taoists sit and forget self until the *wu* of self and the *wu* of the divine are one. *Advaita Vedanta* advises seekers to strip themselves of the five scabbards of self to reveal non-duality with the divine without qualities. *Samkhyans* observe the world to find they are not part of it; they are nothing and nothing is awake.

The religions of the West, while not focusing upon the ontological status of individual self, staunchly hold to the seeker's prerequisite need to be stripped, actually making effort to do the stripping, of all individual attributes, to become empty of personal *élan* (life) utterly. In the Gospels and for Saint John of the Cross, this personal rarefying is called *kenosis.* Meister Eckhart called it *heiligung.* The Jewish Kaballists added their voice with *bitul.* Islamic Sufis brought their language of *al fanaa,* dying before you die. These are nine traditions giving voice to a single notion: the seeker empties themselves of their self in order to unify with the most rare divine as Nothing.

So, What is Nihilitive Knowing?

We have, in a paced manner, walked toward the goal of answering this most important question: what, exactly, is Nihilitive Knowing? We will approach the answer along three different pathways: the psychic nature of knowing, the personhood of knowing, and the content of knowing. Each of these has been cleared and leveled by our discussion thus far, and as we shall discuss now, each pathway will lead us over a cliff! Nihilitive Knowing is over the cliff edge of the psyche, over the cliff edge of the human being, and over the cliff edge of content.

Now, knowing is normally considered, at base, a psychic phenomenon. As we laid out the structure of ways of knowing, we used the term "psyche" to refer to the realm consisting of the conscious and unconscious. The conscious is our waking experience of daily perceptions, emotions and goal seeking; the unconscious is the "un," the Freudian atavistic force zone, or the space of right brain activity and parallel processing of information. We referred to the unconscious as "meta-conceptual" in that thought has no *élan* (life) therein. However, certain people have the capacity to stay awake in the meta-conceptual as witnessed by the example of August Kekulé in his reverie on the self-encircled snake.

In completing our structure of knowing, we posited a realm that lies beyond the psyche itself, the "super-psyche."We discovered that religious traditions from around the world exhibit a great facility at creating ways of staying awake in this realm, from concentration on the constituents of selfhood to sitting-forgetting, from the debating style of *apophasis* to the self-abasement of dying before you die. Each of these traditions was expressing what we termed the "rarefying principle": in their most stripped away and ultimate nature, the divine and the human are both Nothing.

Thus, the first pathway toward our most important question, the realm of the super-psyche, i.e., over the cliff edge of the psyche, is the realm of Nothing. In the meeting of Nothing and Nothing, Nihilitive Knowing occurs. It is the simple fact of being awake in the super-psyche.

Secondly, we outlined three sorts of knowing in our topology. Objective knowing is the pursued goal of scientists regarding the world around them that achieves validity by being verified in that outside world. Subjective knowing resides as epistemic feeling inside an individual that finds its validity within as an internal confirmation. Intersubjective knowing is social in nature and finds its validity in the relationships of a group who agree on what Husserl termed a shared *heimwelt* (homeworld).

So, we can now ask for the final time our topological question on sorts of knowing: Where does Nihilitive Knowing reside? First, Nihilitive Knowing occurs for a certain person. Yet, of necessity, Nihilitive Knowing only arises when the person is without thought, without psychic élan. Nihilitive Knowing only arises when the person is Nothing. Remember that Nothing cannot be experienced, since by "experience" we presuppose a human being awake in the conscious psyche. The term "experience" cannot be used regarding Nihilitive Knowing. Perhaps, we could coin a new, more exact term: "inperience"!

Thus, the "where" question of our topology also leads us over a cliff edge. Nihilitive Knowing finds no place to reside, neither objective to an individual human, nor within an individual, nor in the relationship of individuals. For it only occurs when the individual is negated unto Nothing, thus making the topology of objective, subjective, and social inept to find it a place. Nihilitive Knowing lies beyond the cliff edge of "person."

Furthermore, we can now take a more expansive and conclusive look at the reason that religious speak. The speaker, having met the divine Nothing while residing as Nothing, becomes non-different from the divine. Earlier, we presented research on the meta-conceptual effects of speaking that have been termed "neural resonance" or "mirroring other people's neuroacoustic properties" or "brain mirroring." So, regarding the speaking of Nihilitive Knowing, the speaker can be said to attract the listener to mirror the consciousness of the speaker and thus be wooed to meet with the divine Nothing as well.

It may be helpful in appreciating this somewhat subtle point to call upon the language of Taoism. Taoism, we contend, has perfected, beyond any other religion of the world, the metaphor of the mirror to teach us the nature of the mind in its most rarefied state. Taoism calls one who has met with *wu* (Nothing) The Perfect Man. The mind of such a one is said to be the mirror of the entire world. Let us once again turn to Zhuangzi for a classic statement on mirror mind:

Hold on to all that you have received
But do not think that you have gotten anything
Be empty, that is all. The Perfect Man uses his mind like a mirror
Going after nothing, welcoming nothing, responding but not storing

Lee Yearley does a good job of voicing the connotations of Zhuangzi by stating that the human mind reflects emotional states such as joy, grief, and hate as well as cognitive states like imagination, dream, and thought. "It watches those states rise and fall and thereby remains unaffected by them... It mirrors despair or hope but is not itself despairing or hopeful."

The Perfect Man, one who has greeted Nothing with Nothing, enacts their own emotions, thoughts, and desires without disturbing the reflective élan of the placid mind. Further, they "respond without storing", i.e., they speak without the mind itself being ruffled. It remains "empty, that is all." The religious, in the midst of their speaking, reflect back to the listener who the listener is in their most rarefied form as Nothing.

Thus, speaking Nihilitive Knowledge has great effect on the listener. The religious speak with great impact, however the pith of communication of Nihilitive Knowing is without content. Recall our discussion before regarding Nothing and nothingness. If the divine is beyond concept of any sort, all talk about the divine, even words of negation, is talk of the divine as nothingness, i.e., divine within the realm of concept. When the religious speak their Nihilitive Knowing, their **words** are about nothingness. The content of their speaking is nothingness. However, the religious person who is awake in the super-psyche, who is awake as Nothing, has their effect in speaking meta-conceptually, super-psychically, over the cliff edge of content.

Thus, we give this conclusive answer to our important question: What, exactly, is Nihilitive Knowing? Nihilitive Knowing has its élan over the cliff edge of psyche, over the cliff edge of person, over the cliff edge of knowing content. Nihilitive Knowing arises where the Nothing of the human being and the Nothing of the divine greet. It is the simple fact of being awake in the super-psyche. Speaking Nihilitive knowledge mirrors the listener and speaker, pulling the listener into the Nothing of the speaker.

Now, there is one final aspect of our investigation of Nihilitive Knowing that we will pursue as a coda to this discussion. We have been focusing intently on the nature of human and divine Nothing. How does this focus add to the effective range of Nihilology as a heuristic?

Let us reveal one more regimen of language that we have undertaken for the purpose of clear understanding for our readers. Throughout our book so far we have rigorously approached the usage of the term "Nothing." We have been very disciplined in refraining from the addiction to all-too-easy jokes about the multi-valent connotations of the term which abound in the literature on Nothing. (Though, we heartily agree, they are often very funny.) We have also not resorted to self-contradictory language and facile word shuffles to hint at the nature of Nothing, for, in the end, this sort of talk is intellectually lazy and leaves things unclear. What we have done, without exception, is to use '"Nothing" as declarative, absent any undercutting or paradoxical negation.

In doing so, and we hope to have been successful in this regimen, we have slowly and steadily built within the conscious psyche of our readers a Nothing concept that is clear and declarative. Recall the statement we headlined as an epigraph to our book: "In this phase of life, nothing is more essential to us than academic career, professional respect and reasoned science." When readers came to the first page, fully educated and alert, they probably heard this headline to mean something like: "Having come to a place of maturity, our professional success, collegial status and research findings are important above all else." And this meaning is as it should be for readers who have learned to hear

"Nothing" as our culture has taught us. Yet now, our readers who have stayed loyal to our discussion for 82 pages, may hear the headline thus, "Having come this far in the book, Nothing tugs at us so; even our job, our acclaim and our publications bear on us less."

Going one final step, we have attempted also to plant in readers' meta-conceptual psyches a simple sense of Nothing's nature as one of verity and gravity, a Nothing that resides in our readers as true with a pull to it. And, what is most important and difficult to effect, the Nothing sense residing in the realm of the meta-conceptual has remained free of any inherent characteristics and has no ideational handles at all. It resides as a verity with a pull to it.

Soon, we will move on to a critique of popularizers of science. This conscious declarative and meta-conscious verity/gravity will serve as yet another heuristic toward understanding their writings via the theme of Nothing. We can test their arguments by demanding congruence with the declarative concept in our conscious psyche and listening for resonance with the true pull in our meta-conceptual psyche. This is an exacting test, and it is our position that it is one the popularizers fail to pass.

Our New Model of Insight

Let us conclude this section on Nihilitive Knowing by clearly elucidating, as we have promised, a new and more penetrating model of scientific insight. It will be noted below in our critique of the popularizers of science that, to a cosmologist and astrophysicist, they hold to the reality of insight and to its central force in scientific discovery. Yet, they offer no psychologically sophisticated model of how insight comes to be. With all that we have discussed serving as foundational perspective, it is now possible for We introduced before how scientists reflect on a problem intensely over a long period of time. With the fatigue that sets in from no solution, no breakthrough, they often turn to the use of images to propel them farther along toward a solution (Remember Crick & Watson or Steven Hawking). We termed images "trans-heuristics," aids for the scientist to free themselves from the drag of conception—from paradigmatic constraint—thus to attain escape velocity.

A long period of intense investigation is often followed by a period of cognitive rest, as noted in the work of Lonergan, Wallas, and others, wherein the scientist may muse on other topics, enter into reverie, or simply sleep. This period of rest ends when the scientist's thinking mind suddenly reawakens and the first activity, if they are lucky, results in a burst of energetic thought that solves the long-sought problem. This energy of insight gives birth to the sucking in of breath and the elation of a-ha! us to offer this sophisticated model.

This is scientific insight, according to popularizer scientists, and many researchers into the phenomenon. Yet, what is occurring during the rest of cognition? Where is the élan of the scientist during cognitive absence?

We have discussed extensively the realms available to human beings beyond the conscious concept-making psyche. For many scientists and writers about insight, this meta-conceptual is simply the "un," being of the nature of unawareness and remaining unknown. With a bit more probing into the work of cognitive psychologists, neuropsychologists and neurotheologians, we were able to shine a light into that "un." We saw that the meta-conceptual is actually a part of the psyche called the unconscious, wherein instincts, right brain activity, and parallel processing of information occur. Further, from the religions of the world, we learned of the super-psyche, a realm not only over the cliff of concept, but beyond the psyche itself. It is the realm of Nothing.

Going one step farther, we learned that certain people, with the aid of specific modes of training, gain the facility to stay awake in the meta-conceptual. Psychoanalysis, lucid dreaming practice, and *träumerei* are examples of such training. Others, taking on the disciplines of the world's religions, develop the facility to stay awake in the super-psyche. Sitting-forgetting, *koan* study, and *dhikr* are among these disciplines. In the super-psyche, the divine is encountered when the nature of the divine and the nature of the person are rarefied to Nothing. Nothing greeting Nothing; this is Nihilitive Knowing.

Now, when the trained or disciplined emerge from awakeness in the meta-conceptual or the super-psyche, blooming once again into the realm of cognition, their thoughts are empowered. The empowered thought, for the scientist, may result in the eureka! of discovery. Yet, we saw that religions also acknowledge this empowerment of emergence. Recall this teaching from Jewish Kabbalah, "When one attains the level of gazing at *ayin* [divine as Nothing], one's intellect is annihilated. Afterward, when one returns to the intellect, it is filled with emanation."

We hold that insights are actually outsights. They occur upon the *exit* from the meta-conceptual or super-psychic. And it must be said that, for those of religion, these outsights are merely tertiary events to the divine encounter as Nothing. For those of religion, it is Nothing that is the goal and to remain in the condition of Nihilitive Knowing that is sought after. The exit, accompanied by outsight, is beside the point. For the religious, outsight is actually a somewhat immature byproduct.

Yet, for scientists, the outsight is the goal in that the outsight brings to light the long-sought-after answer to their beloved problem. For the scientist who is unprepared and undisciplined to stay awake in the meta-conceptual or super-psyche, there is no élan in the realm of instinctual forces and self-encircled snakes and especially no élan in the realm of Nothing. For them there are only four stages to insight: intensive thought, conceptual rest, eureka!, and verification. However, it is our contention that this four-staged model is superficial, in the root sense of that term. This model, held by most scientists and writers on scientific discovery, describes only one side of the boundary between concept and the meta-conceptual. It leaves the other side unexplored. Most scientists are ignorant of the meta-conceptual, ignorant of Nothing.

Thus, it is our contention that scientific discovery, usually termed insight, is actually a five-staged process: intensive thought/employment of images, conceptual rest, meta-concept/Nihilitive Knowing, a-ha! outsight, and verification. This five-staged model offers a psychologically sophisticated and analytically strong model of scientific discovery. It boldly interprets, using our findings from the religions of the world as well as the history of science, the essential feature of the insight process that most writers have left dark or blithely relegated to the realm of the "un."

Soon, we will enter into a dialogue with the popularizers of science. We will employ this five-staged model to reveal the unsatisfactory nature of their understanding of scientific discovery. We will see also that outsight works as a key heuristic concept to critique the popularizers regarding the epistemology of Nothing.

In a brief foreshadowing, we hold that, since essentially all popularizers of science allow for, and believe in, the efficacy of outsight in furthering their understanding of the universe, they must, by force of rational argument, also allow for the actuality of the meta-conceptual and super-psyche. Resulting from our presentation of the five-staged outsight model, we conclude that the popularizers take the tertiary benefits of these realms, yet deny their existence! We will argue that the popularizers do indeed delve into the realm of meta-concept and super-psyche, but without awakeness therein. We will suggest that their lack of awakeness in the meta-conceptual and super-psyche results from lack of practice and lack of discipline. (They could benefit from the teachings of the world religions on ways to stay awake.)

(5)
Nausea and the Übermensch

We have dedicated much time and energy to elucidating the history of Nothing, how Nothing is taught in the world's religions, and how Nihilitive knowing comes about. And now, before entering into a discussion of how to use the heuristic tools of Nihilology, we must make a revelation: there has been an important reflection on the nature of Nothing and nothingness in the world of western philosophy that we have totally ignored. We have turned our gaze intentionally away from these reflections because these western philosophers have founded their views in naughting the religious teachings that we have elaborated on thus far. These minds could not be fathomed without first understanding their foil, religions, from the vantage point of our empathetic analysis. So now we can stop averting our gaze and enter earnestly into contending with those thinkers whom we have been ignoring.

As was the case with our discussion of religions, we do not intend to provide a wide-ranging inclusive history of western philosophy. We leave that to other scholars. We will focus in on Existentialism and Nihilism for, indeed, these eddies of philosophical reflection were agile—and arguably ruthless—in their use of our central notions of

Nothing and nothingness. And we will allow the generational voices of these philosophies, namely Jean-Paul Sartre and Friedrich Nietzsche, to speak for the schools as a whole.

Our intent in this section is to build toward an understanding of the experience of nothingness as nausea/anguish according to Sartre and the *übermensch* in the creative imagination of Nietzsche. This will allow us yet another set of effective heuristic tools in the critique of popularizer scientists.

Now, as we introduced before, Existentialism and Nihilism worked hard to naught the central premises of the religious worldview as they saw them expressed in Europe at the end of the nineteenth century and the beginning of the twentieth century. For example, Existentialism and Nihilism go straight for the heart of the matter and deny God. Sartre, in his magnum opus *Being and Nothingness: An Essay on Phenomenological Ontology*, explains that human beings have long believed in trans-sensory beings and reality, and as a direct consequence, believed the daily appearances of the world to be somehow mistakes with no true being. "But if we once get away from what Nietzsche called the illusion of worlds-behind-the-scene... then the appearance becomes full positivity; its essence is an appearing which is no longer opposed to being but on the contrary is the measure of it. For the being of an existent is exactly what it appears."

For Sartre and Nietzsche, there is no real hidden under, above, or behind the daily world of common perception and reflective thought. This hidden is the realm normally referred to in philosophical circles as the noumenal, and according to Sartre and Nietzsche, it is the fantasy of such philosophers as Emanuel Kant as well as the priests of Christianity who believe in the ultimate noumenal reality: God. Sartre and Nietzsche only allow, and stringently so, for the reality of phenomena, and for this most basic tenet they are often referred to as phenomenologists.

In the same passage, Sartre deepens his argument with an illustration from the sciences. Force is not a metaphysical energy of a mysterious sort that hides behind its effects on objects. For example, electric current does not have a secret hidden side. It is simply the totality of the physical-chemical actions which make it manifest. It doesn't point beyond to an unseen reality; "it indicates only itself."

(It should be noted that, later, we will encounter echoes of this classic argument voiced from the mouths of the popularizer scientists. It will be soundly critiqued with the heuristic tools of Nihilology.)

Now Nietzsche and Sartre didn't restrict themselves only to scholarly repudiations of God on the grounds of their phenomenological arguments; they also employed a pointed, heavy-hearted sarcasm. In his book *Ecce Homo: How One Becomes What One Is,* Nietzsche refers for his humor to Stendhal, the nineteenth-century French fiction writer, "[Stendhal] took away from me the best atheist joke that precisely I could have made: "God's only excuse is that he does not exist."

And Sartre tells a wonderful story from his childhood to make fun of the Christian God. He had burned a hole in a throw rug in the bathroom where he was hiding to play with matches. In the act of trying to cover up the burn, he suddenly experienced God watching him from inside his own head. He felt nakedly exposed, yet instead of bowing in shame, he threw a childish tantrum at this intrusive God. "God damn it! God damn it!" he shouted. Sartre wasn't actually asking God to condemn himself. He could just as effectively have shouted "Merde! Merde!" Rather he was simply yelling out the first blasphemous, dirty words he could come up with to shoo this easily embarrassed God out of his head. And it worked. According to Sartre, "He never looked at me again."

Further, Nietzsche in particular was quite caustic about the history of Christianity. He was well-known for indicting the tradition as being born out of people's disgust with life and for describing the Eucharistic imbibing of the blood of Christ as ghoulish.

We should note that Nietzsche was born into a long line of ministers. He was very religious as a child and was expected to become a minister himself. As we shall see, it seems he never outgrew his religious role. Indeed, it expands to a somewhat megalomaniacal persona as the übermensch.

In his most influential book, *The Will to Power,* Nietzsche blamed Christianity for the malaise of nineteenth-century Europe. He envisioned a massive pendulum that for two thousand years had been swinging out to the extreme of its arc by overvaluing the human, then surging to its opposite extreme and crashing an entire continent of people into pessimism and self-loathing. "The sense of truthfulness, developed highly by Christianity, is nauseated by the falseness and mendaciousness of all Christian interpretations of the world and of history; rebound from 'God is truth' to the fanatical faith 'All is false'..."

And Sartre, while not as caustic as Nietzsche regarding the Christian faith, was just as vehement in preaching the notion that God was beside the point (Much like human beings who, as we shall discuss next, he says are *de trop*—superfluous). In his famous speech "Existentialism is a Humanism," which he later regretted delivering, probably because he defended his philosophy too clearly and simply, Sartre declared that Existentialism would not waste its time arguing the non-existence of God as his being or not being was irrelevant in the end. The real problem for people of Europe, in his view, was to rediscover their true nature and to accept that they cannot escape themselves, even if they discover a "valid proof of the existence of God."

(This section of Sartre's speech is echoed, down to the phraseology, in a book by one of the popularizers of science whom we critique below, Lawrence Krauss. It would seem Krauss cribbed Sartre for this philosophical position.)

With God dead, or at minimum beside the point, both Sartre and Nietzsche envisioned an enhanced role for humans in the world. In his work *Notebooks for an Ethics* Sartre embraced the possibility for humanity to become "the heir of the mission of the dead God." This mission, according to Sartre, was to draw meaning out of a godless world.

Nietzsche speaks of an even grander role for human beings in his fantastic *Thus Spoke Zarathustra*. He declares since God lies dead in his tomb, it is the *übermenschen* who are resurrected at the "great noon" of humanity and become lords of the people.

Thus, we see Existentialism and Nihilism ridding the universe of any hidden, behind-the-scenes realities, including God. Theirs is a world of daily phenomena, ruled by human beings who must create and guide their own lives. It is within this world that we now turn to examining Nothing and nothingness.

As we noted before, the writings of Jean-Paul Sartre will serve to highlight the philosophy of Existentialism. Further, it must be noted here that the person—or shall we say persona—of Sartre's devoted creation was always noticeably present in his work. Thus, we contend, the most penetrating and sensitive interpretations of Existentialism always incorporate this persona. (This is the case, to no lesser extent, in the

Nihilism of Nietzsche and, as we shall discuss below, in the cosmological writings of the popularizers of science.)

It is an oft-told story that young Jean-Paul was doted upon as a child; his family believed that he was a beautiful boy. The adult Jean-Paul internalized this belief, exhibiting a narcissistic hue throughout his very public life. He promoted a cult following in Europe beginning in the mid-1940s, and his fame bloomed in the US in the alienated 60s. He was granted the Nobel Prize in Literature in 1964, but he denied the award in order to be seen and read in the world as an institution-free individual. He paraded as the cynosure of the Left Bank café culture. During one period he was deemed to be one of the most famous people in the world. There was no room for a divine being nor even a realm outside the phenomenal for Sartre; perhaps it would have meant sharing the spotlight! As he declared of himself, "I want to be the one who knows the most things. I've got a golden brain."

Now, consistent with these thoughts on his persona, we must see right off that Sartre was not promoting a metaphysical view of human life in his Existentialism. He attempted no explanation of the way things are in the universe, nor of the **why** of the way things are. His philosophy is, rather, an ontology; he describes the **how** of the universe. (Sartre is paralleled by the popularizers of science who hold that there is no **why** to the universe.)

And, regarding the main theme of our book, his decidedly non-metaphysical view of the universe distinctly delimits the **where** of Sartre's Nothing. As he declares in *Being and Nothingness*, borrowing the phrase of the influential Martin Heidegger, "the world and outside of that nothing." Nothing, for Sartre, is not a metaphysical out of which human reality emerges. There is no such metaphysical. (Let us postpone for a few paragraphs an explicit discussion of what Sartre intends by "nothing" For now, simply note that his language is quite inconsistent in this regard, with "nothing" carrying at least three different meanings.

Let us go further in continuing to point out the gulf that exists between the Sartre's philosophy and teachings of the world's religions regarding the nature of the universe. Recall that the Latin church of the third century entered into a contest with the Gnostic movement of the Eastern Church over the metaphysical state of the world before creation. The Latin Church held, and held sway historically, that God created the universe *ex nihilo,* out of Nothing. Whereas the Gnostics proposed that God made use of a lesser divine to create the universe out of *tohu va vohu,* a stuff without form. Now, Sartre (and this applies to Nietzsche as well) was not interested in whether the world came to be out of Nothing or a pre-existent stuff. Stuff/Nothing; these were simply not of interest. He witnessed the phenomena of created existence, and that was enough for him. His was a human-centered world and, for the persona Sartre, a self-centered one.

Sartre was not a mystic, nor did he accept the canonical reports of religious mystics about noumenal experience. To those mystics who claim transcendental encounters or divine sanctioned mission, Sartre counsels that they are reporting only human thoughts and human aesthetic experiences with the intent of avoiding human responsibility for their lives. Whereas the religious takes on a disciplined practice toward rarefying their self unto Nothing with the hope of greeting the divine as Nothing, Sartre teaches that the highest human practice is to seek a self who can never be found! (Sartre was the popular star of the life around him. There is no possible way of him surrendering this laud of the public eye by giving up his individuality at the feet of a divine.)

Hazel E. Barnes, translator of *Being and Nothingness*, is insightful in this regard. She tells us that, for Sartre, "acceptance of one's absolute freedom is the only existence commensurate with an honest desire to exist fully as man. But the recognition comes not in ecstasy but in anguish. It is not a merging with a higher power but a realization of one's isolation..."

However, our readers should not think that Sartre had no intensive or life-changing experiences. It is just that he could not entertain the possibility that his experiences were of a trans-human origin or locale. For example, he reported a day of seeing into the very nature of being. Yet, this insight revealed life in its phenomenal nature, revealed the superfluous nature of being alive, and revealed the foundational tone of nauseating human existence. Here are two more poignant examples from his novel *Nausea*, wherein Sartre embodies himself in the lead character Antoine Roquentin:

Then I had this vision... there it was, clear as day:
existence had suddenly unveiled itself
the diversity of things, their individuality, were only an appearance, a veneer
This veneer had melted, leaving soft, monstrous masses, all in disorder
naked, in a frightful, obscene nakedness

I'm suffocating: existence penetrates me everywhere,
through the eyes, the nose, the mouth
And suddenly, suddenly the veil is torn away
I have understood, I have *seen*
I can't say I feel relieved or satisfied; just the opposite, I am crushed

Furthermore, recall that we went into rather great depth to elucidate a world-wide theme in theology termed the *via negativa* that served to deny the rational mind's capacity to encapsulate the divine as Nothing. We called upon the teachings of Buddhism because they are clear in this regard: the **idea** about Nothing is not Nothing itself. Ideas form around Nothing. Ideas are not about Nothing at all.

Now, in a certain sense of the phrase, the existential philosophical approach of Sartre, and Nietzsche as well, was a sort of *via negativa*. Yet, his mission was a denial of moral guidelines and of metaphysical reality. He was not concerned with the ability of concepts to encompass the divine. For indeed, as Sartre and Nietzsche joked, God's one essential downfall is that he doesn't exist.

Now, we have spent many words in this discussion to differentiate the Existentialists, specifically Sartre's voice, from the metaphysical approach to reality as witnessed in the world's religious traditions. Yet, readers have certainly noticed that many large statements by Sartre and Nietzsche have a metaphysical ring to them. They both employ transcendental sounding terms like "existence," and "eternal," and "being." Nonetheless, we must see that they were simply making a cunning use of vocabulary from the religious traditions of Europe, as well as the philosophical traditions that recognized the divine, as tools to undermine the teachings of those very same traditions!

For instance, Sartre often proclaimed "existence precedes essence," a bold reversal of the religious teaching on the omniscient and omnipotent creator. However, this usurpation of language was not intended to attempt to understand the true means of creation, nor to put forth a more profound interpretation of how humans came to be. Note Sartre's casual and poetic language describing human creation, "[A human] surges up in the world," and "[human birth is] that leap toward existence."

Rather, Sartre was attempting to explain the psycho-emotive experience humans need to create their nature... by themselves... within their lives. Let us quote a rare simple statement from his speech "Existentialism is a Humanism" that makes his position plain:

What do we mean by saying that existence precedes essence?
We mean that man first of all exists, encounters himself, surges up in the world
and defines himself afterwards... Thus, there is no human nature
because there is no God to have a conception of it. Man simply is
Not that he is simply what he conceives himself to be
but he is what he wills, and as he conceives himself after already existing
as he wills to be after that leap towards existence
Man is nothing else but that which he makes of himself
That is the first principle of existentialism.

And Nietzsche also usurped language to undermine the teachings of religious tradition in declaring his notion of eternal recurrence, wherein humans return an infinite number of times to live exactly the same life "in what is greatest as in what is smallest." For example, the phrase "eternal recurrence" translates *die ewige wiederkunft* that is traditionally used to reference the second coming of Christ. Also, this passage from Nietzsche's allegorical parable *Thus Spoke Zarathustra* where the hero Zarathustra is addressing himself uses the same language:

O Zarathustra... behold, you are the teacher of the eternal recurrence
We have already existed an eternal number of times
and all these years are alike in what is greatest as in what is smallest
We ourselves are alike in every great year
I come again with this sun, with this earth, with this eagle, with this serpent
not to a new life or a better life or a similar life
I come back eternally to this same, selfsame life
to teach again the eternal recurrence of all things
to proclaim again the overman [übermensch] to men.

We will take a deeper look later at Nietzsche's Nihilism and his übermensch. For now, let it suffice to say that the übermensch was a prophet born out of humanity, that the übermensch superseded the role of Christ, and that Nietzsche envisioned himself as the first of these prophets.

Now, for all Nietzsche's grandiose language here, his intent in teaching eternal recurrence was not a metaphysic *per se*. The übermensch, for instance, does not come to the world in order to free himself or his disciples from the rounds of life; he is not a Hindu *avatar* descending into the human realm. Secondly, note that Nietzsche, neither in his own voice nor in that of his fictional persona, does not devote a single sentence to justifying why the world and humans return, nor does he explain the mechanism of how such a recurrence would work. (Does each individual return to the self-same womb at the moment of death? How does one coordinate the readiness of our womb to return with our readiness for the womb?) Thirdly, what justification would the übermensch have for teaching humanity when no learning or evolution is possible?

No, Nietzsche was not offering us a badly argued metaphysic. He was, in a philosopher's sleight of hand, employing metaphysical language to point out to human beings that we must choose our lives fully, that we need to make a choice in the present to avow the life we are leading, that we must embrace our lives as they are. No etiology nor teleology, no ultimate cause nor end.

Now, having accepted that Sartre—and Nietzsche as well—focused his thoughts on the elaboration of a phenomenological ontology, repudiating any reality to the noumenal realm and not intending to provide a metaphysical philosophy whatsoever, what did he propose as the nature of existence (being-in-itself) and human existence (being-for-itself)?

Being-in-itself simply **is**. It is not derived from anything else and the question of its possibility or impossibility is irrelevant. Being-in-itself is uncreated, without reason, without connection. It is for eternity *de trop* or superfluous. Being-for-itself, the human being, is conscious of its own actuality, what Sartre called "facticity." Yet human beings sense that they exist for no particular reason. "The being of human reality is... a totality which it is without being able to be it." The human is also *de trop*.

These are essential positions of Sartre on the nature of existing. Being has no purpose. There is no built-in meaning for existence and thus no meaning for human beings as well. (*De trop* literally translates as "too much." This literal meaning fits well with our slang usage, as in "Life is just too much.")

It is important to remember that Sartre is a phenomenologist. He lends reality only to human experience and pursues a philosophy of analyzing that experience. And when he looks at the human being he beholds a life of utter and unresolvable frustration, what he termed a "futile passion." Being-for-itself is drawn without relent to unify with being-in-itself, meaning we as humans seek the factual nature of our existence. Yet, this unity is never found, in fact cannot be found because the self is by its very nature at a distance. Sartre portrays this futility as the haunting of human life as we question our being as long as life withstands this predicament.

In order to make clear the intended sense of this existential axiom, we present this wonderful line delivered by the character Estelle in Sartre's eccentric play *No Exit*.

> I feel so queer. When I can't see myself I begin to wonder
> if I really and truly exist

I've six big mirrors in my bedroom
When I talked to people I always made sure there was one nearby
in which I could see myself

Estelle only felt real when she could see herself in a mirror, when she was an objective observer of herself. This is the theatrical equivalent of Sartre's philosophical statement, "The being of human reality is... a totality which it is without being able to be it." The fact of humans, according to Sartre, is that we are at a distance from true being and doomed to be ever absent no exit. (This is the pith of Sartre on what we now term alienation and is the key to his idol status among the US counterculture of the 1960s.)

The second part of this axiom is that the very nature of our consciousness includes by necessity the doubt as to its reality. Sartre's Existentialism paints a world that is only real on the level of phenomena, which does not allow us to rely on some eternal underneath or behind, and at the same time, we cannot rely on the reality of the phenomena. This is the doom of our "futile passion."

Now, our readers may have noticed that we have seldom allowed Sartre to extensively speak his philosophy in his own words. We have re-phrased, explained, and re-stated Sartre, especially in his passages about being. This labor on our part is demanded by the often opaque, chaotic quality of Sartre's technical language. Many interpreters have noted this shortcoming. Wyatt and Schnelbach don't hold back their criticism on their website called "The Existential Primer." They observe that Sartre was internally impelled to out-philosophize the revered voices of European philosophical history, and this led him to forgo ethics in order to win debates. He demonstrated a personal profound lack of respect for intentions and even the truth. Sartre felt no remorse in lying or intentionally distorting the theories of other philosophers.

Also, Walter Kaufmann, the wide ranging translator and interpreter of Sartre, had this biting perspective, "The pity of it is that Sartre clothes his analysis in spurious dialectic: he speaks of 'the nothing' like Heidegger, takes 'in-itself" and 'for-itself' from Hegel, and above all plays on the word 'being' in a way that veils his meaning from most readers... "

Sartre saw himself very much a member of the European philosophic community (and strove to be the cynosure here as well). He took great effort to include himself with, and to distinguish himself from, René Descartes (1596-1650), Immanuel Kant (1724-1804), Georg Hegel (1770-1831), Edmund Husserl (1859-1938), Martin Heidegger (1889-1976), and other lights of western philosophical thought. This was a trap for Sartre in that it necessitated he speak in the categories of the tradition, that is, being, non-being, essence, in-itself, and so forth, but his language became overly complex, slightly out of control, and oftentimes "spurious."

Were he able to extricate himself from the need to join and overcome his colleagues, he could have employed much simpler language, for indeed his points about the human experience are straight forward. The human dilemma is that we sense our very existence is superfluous (*de trop*); there is no floor beneath us and no divine guidance for our days on earth. This dilemma would be fairly simple to write about if Sartre were free of the burden of unnecessary homage to and crusading against his philosophic tradition. Ultimately, he made himself a victim of his own aspiration toward beating the great ones and his need for the fame that winning entails.

However, while Sartre's driven philosophical debate often led to unnecessary complexity and misdirection, he was indeed a genius in the expression of the psycho-emotive tone of the Existential dilemma of human life. It is in the mirror of his works of fiction—the play "No Exit", the novel *Nausea,* and the memoir *The Words*— that we are offered a true reflection of Sartre's vision. (Yes, his memoir is of the fiction genre because, in the book, Sartre curved and skewed his early life to fit with the famed persona of his later years.) Let us delve specifically into his novel *Nausea* as a means of appreciating Sartre's genius in this regard. The first person singular in the following quotes is Antoine Roquentin, the voice of Sartre as a fictionalized character:

I was there, motionless and icy, plunged in a horrible ecstasy
I understood the Nausea, I possessed it
The essential thing is contingency (*de trop*)
When you realize that, it turns your heart upside down
and everything begins to float... here is Nausea

Sartre employs the term nausea (*la nausée*) to describe an individual's recognition of the pure contingency of the universe and their personal life. This recognition occurs at the meeting place of mind and body. Existence is "too much," and we are sickened by the seeing.

Things are bad!... I was surrounded
seized by a slow, coloured mist, and a whirlpool of lights in the smoke
I no longer knew where I was
I saw the colours spin slowly around me, I wanted to vomit
And since that time, the Nausea has not left me; it holds me

The color tone of such a contingent life is gray. Gray fills the air as a fog. (We are tempted to use the term "black," but that hue is too definitive, too reliably real.) So, not only did Sartre think the nausea in his mind and feel the nausea in his body, he also saw it with his eyes as a physical presence in the air!

The Nausea is not inside me
I feel it *out there* in the wall
in the suspenders [of the waiter], everywhere around me
It makes itself one with the café
I am the one who is within *it*

With this disturbing realization, Sartre further loses controlling boundaries for the nausea. It is the very constituent nature of the environment in which he lives. Nausea surrounds him. It is everywhere.

The Nausea has not left me
I don't believe it will leave me soon
but I no longer have to bear it
it is no longer an illness or a passing fit: it is I

These passages from Sartre's fiction provide us with a distinct aesthetic understanding —psycho-emotive tone—of the Existential dilemma of the human being. There is no divine guidance nor dependable ethical guideline for living. Human life is beside the point, superfluous (*de trop*); it is nausea. Yet, nausea, according to Sartre, is not simply the psycho-physical upset we feel as humans in the world. Nausea not only surrounds the human being as a visually sensible presence. Nausea **is** the human being. Nausea are us.

Let us then return to Sartre's philosophical writings armed with this clear sense of the human being as nausea. Note particularly these two statements from his tome *Being and Nothingness*:

The necessary condition for our saying *not*
is that non-being be a perpetual presence in us and outside of us
that nothingness haunts being

Yet the for-itself *is*. It *is*
But this being is constituted as human reality
inasmuch as this being is nothing but the original project of its own nothingness
Human reality is being in so far as within its being and for its being
it is the unique foundation of nothingness at the heart of being

The first statement consists of two phrases in apposition. Note the words "that" are used in parallel. This syntax demands that the subjects of each phrase are interchangeable or synonymous according to context. Thus, the first phrase could also read, "nothingness be a perpetual presence in us and outside us." This is the exact teaching Sartre created in his fictional work quoted before; only in that aesthetic context he made use of the term "nausea."

The second philosophical statement can be more simply stated as, "Yet the for-itself... (human... being) is... its own nothingness.... the unique foundation of nothingness at the heart of being [in-itself]. As Sartre said in his fictional work, "The Nausea... is no longer an illness or a passing fit: it is I."

The main point is that Sartre used "nausea" and "nothingness" as virtual synonyms, one in the fictional context and one in the more debate-oriented philosophical context. The human being, being-for-itself, is constituted of these. Nausea/nothingness is the very essence of the human being and important to the overall analysis of our book toward the building of Nihilology. This is the first meaning of the Existential nothingness: nausea.

Furthermore, there is a second meaning of Existential nothingness: anguish, what we commonly call anxiety. There is one straightforward and telling metaphor toward understanding what is intended here. Sartre was drafted into the French military following the German attacks in 1939, being assigned to a meteorological unit that sent up weather balloons. Only a year into his conscription he was captured and imprisoned by the Germans. It was in this prison camp that Sartre began the composition of his most widely known philosophical work: *Being and Nothingness*. This context, ironically, revealed to him the Existential ult-fact: human beings are, by nature, utterly free!

Sartre is indeed a master of odd paradox, "... man is condemned to be free.... thrown into this world... responsible for everything he does." Just as soldier-Sartre was thrown into prison camp so are we as humans thrown into life, not so much against our will rather without any say or guidance whatsoever. (Recall that we inexplicably "surge up in the world.") And once a prisoner, we are personally responsible for our every act. We are forced to freely choose! And this dilemma of imprisoned freedom of choice we experience as anguish. Anguish is not simply a tertiary experience of freedom, it is the very nature of freedom itself, the defining characteristic of how freedom enters human life. "It is certain that we cannot overcome anguish, for we are anguish... the flight from anguish is only a mode of becoming conscious of anguish."

Recall before we discussed how Sartre held as one of his central tenets in the Existential life that "existence precedes essence." We surge up in the world unformed and must self-create our essence through choices, which we now know constitute human freedom. Once again, let us turn to Sartre's fiction for a clear proclamation of this teaching in a dialog between Garcin and Inez, two main characters in the play "No Exit":

I *dreamt* you say. It was no dream
When I chose the hardest path, I made my choice deliberately
A man is what he wills himself to be

One always dies too soon—or too late
And yet one's whole life is complete at that moment
with a line drawn neatly under it, ready for the summing up
You are—your life, and nothing else

Human beings are defined, absolutely, by their choices. We are, in fact, created by our choices. We enter life aided by no divine underlying essence and exit life with no promise of eternal existence. This is our freedom and we experience this freedom as anguish.

Now to go one step further in this analysis of freedom of choice as synonymous with anguish, Sartre declared in *Being and Nothingness* that humans possess a "nihilating freedom" that is experienced as anguish. Sartre boldly stated, as quoted before, that "man *is* freedom" and "we are anguish." Yet now Sartre adds one more term to our synonym pair of freedom = anguish, creating a synonym trio: freedom = anguish = nihilating.

Let us turn to the analyses of three expert sources to help us make sense of this trio. Hazel E. Barnes, in the preface to her translation of *Being and Nothingness*, states that the essential human experience is the discovery that we are wholly a self in process, yet we must try in vain to escape the "nothingness that is us." The editors of the Stanford Encyclopedia of Philosophy echo this analysis in saying the human being (for-itself) is fluid. "It is the internal negation or nihilation of the in-itself, on which it depends." And finally, these words from a 1996 Paul Vincent Spade lecture, "Non-being exists only on the surface of being.... non-being is something imposed on being-in-itself. We put it there."

Note that in these interpretive statements, we hear yet another trio of synonyms: nihilating = non-being = nothingness. This phenomenon of synonymous usage of central terminology is a habit of Sartre's that we have become used to in reading his work. It is a bad habit, yet we have come to appreciate it as integral to Sartre's aesthetic style of philosophizing.

So, what shall we make of these many synonyms regarding the nature of human life? The human being has no predetermined divine or phenomenal essence. The human being is nothingness. At the same time, and here is another central feature of Sartre's Existential dilemma, the human being, as nothingness, is a forced act of choice. One must continually rebuild oneself by making decisions to embody the self in the deeds of one's life. Thus, humans are in a constant process of re-enacting their very self. Human beings are nothingness enacted!

Now, nothingness, by its very nature as an act, is a negation, or as Sartre says, a "nihilation." But a negation of what? As put forth in the third interpretive quote before, nothingness (non-being) finds its reality "on the surface" of being. (Remember Sartre's maxim, "Existence precedes essence.") Nothingness (non-being) lies within the realm of being. In that human beings are nothingness enacted, we create nothingness, we "put it there." As was discussed earlier, we as humans (being-for-itself) are in a life-long vain struggle to find our self as the fact of existence (being-in-itself), yet as we now have seen, our nature is nothingness (non-being), and thus our struggle serves to undermine (nihilate) the very being we seek. We are by nature a negation of being!

Now, for our further analysis in the service of building the heuristic of Nihilology, we will distill Sartre's double trio of synonyms down to an effective single trio: Freedom = Anguish = Nothingness. The human being is an act of free choice making. The human being experiences this forced choice as anguish. The essence of the human being is nothingness.

At this point, we'll move toward building a keen interpretive tool regarding Sartre's Existentialism. Note that Sartre based his philosophy on free choice, yet he himself made conspicuously cavalier, even stupid, choices regarding his body. He drank alcohol to excess in the café. He took amphetamines to energize his writing then abused sedatives to allow for some sleep. He smoked multiple packs of Gauloises each day; a cigarette in hand became an essential aspect of his public image. (There is a famous cartoon by David Levine in the header of a December 17, 1964 New York Review of Books article where Sartre explains his refusal of the Nobel Prize in Literature. Sartre is depicted as smoking a pipe and holding a lit cigarette at the same time!) In the end, almost totally blind and depleted by two heart attacks, Sartre died of smoking-caused pulmonary edema.

It is our sense that Sartre ignored his health because, in his philosophical vision, no useful knowledge came from the body, rather only from reflective thought. Sartre suffered from a strong bent to mentalize or philosophize the physical. He valued his "golden brain" above all else. Ronald Hayman wrote a penetrating biography titled *Sartre: A Life*. In it he reports Sartre zealously reassured himself that his mental genius was not threatened by any sort of wisdom from his body. He refused fatigue or pain. He detested the time required to wash, shave, brush his teeth, take a bath, or even to *chier*. From a life-long cigarette habit his teeth began to decay, yet he refused to waste time in a dentist's chair.

Sartre's persona, built around the success of the mind, was his *raison d'etre*, yet he was also blinded by his success of the mind. Were he to listen to his body he would have been forced to conclude that his Existential approach to life was flawed in that it fostered suicidal choices.

To return to our main discussion, we have seen that Sartre focused much of his definitional thought regarding nothingness on how it is experienced by the human being. The nothingness of human life as superfluous (*de trop*) is experienced as nausea. The nothingness of forced freedom of choice is experienced as anguish. Thus, we have Sartre equating the physico-emotive experiences of nausea and anguish with nothingness. This philosophical approach is consistent with Sartre's phenomenology, yet it can be quite unclear for many readers just how nausea and anguish are linked. We suggest that the heuristic key to understanding this linkage is vertigo.

The pathology known in medicine as vertigo is a malady of the inner ear that causes extreme dizziness with an associated visual experience that one's physical environment is spinning or melting. The materiality of the environment seems to give way to fluid, unstable ground, not unlike what is reported by people who have lived through an earthquake. The dizziness of the head is directly linked to upset of the stomach which leads to nausea unto vomiting. Also, the fluidity makes one feel as if they are floating and this causes nausea as well.

Further, acute dizziness and spinning causes one to lose a sense of location. Being physically located in a stable environment is an essential aspect of experiencing individuality and self-identity. One must be somewhere in order to be someone. Spinning causes one to fear for one's individuality, doubt one's personhood, and feel anxiety, what we have up to now termed anguish.

Now, there is ample evidence that Sartre suffered from the actual malady of vertigo. For instance, he was released from military service in 1941 due to a condition known as exotropia wherein one eye drifts outward and distorts or even blocks binocular vision. Sartre reported that his condition made it difficult for him to keep upright balance. Also, in the third book of her memoirs titled *The Force of Circumstance*, Simone de Beauvoir documents that Sartre was hospitalized on a 1954 trip to the Soviet Union with acute symptoms of vertigo. It is also known that Sartre was a heavy user of amphetamines. Side effects include dizziness, blurriness of vision, and nausea.

It is our sense that his suffering from the physical condition of vertigo provided the basis for Sartre's genius at expressing the psycho-emotive tone of Existential nothingness in his fictional works. He had the constant presence of vertigo as his teacher. He could contemplate it first-hand in his daily physical life. It made him sensitive to vertigo's nuance.

Once we realize that Sartre suffered from the disease of vertigo, then this fact shows itself throughout his writing. Again, as before, in a more straightforward expression in his fictional works such as *Nausea* he says:

The grey thing appears in the mirror
It is the reflection of my face
If my forehead was surmounted by one of those neutral heads of hair
which are neither chestnut nor blond
my face would be lost in vagueness, it would make me dizzy

I would like to take hold of myself
An acute, vivid sensation would deliver me
I go to sleep with my eyes open
I lose my balance and that wakes me
I find myself straddling a chair, still dazed

[In the cafe] I was surrounded, seized by a slow, coloured mist
and a whirlpool of lights in the smoke
I floated, dazed by luminous fogs dragging me in all directions at once
Then the Nausea seized me, I dropped to a seat
I no longer knew where I was; I saw the colours spin slowly around me
I wanted to vomit

Thus, in our analysis, the physical condition of vertigo was so present to Sartre that it served as an object of reflection that brought philosophical insight: Sartre philosophized his vertigo. However, remember, Sartre committed suicide through his bad choices. And here again, he was also not able to listen to the voice of his body **as body** regarding vertigo either. He made the vertiginous experience, a physico-emotive cluster, into a cluster of concepts. Sartre philosophized nausea, anguish of spinning, and loss of self-orientation from an inner ear problem into the experience of *de trop*. Anguish of spinning and loss of self-orientation from an inner ear malady, Sartre philosophized into the experience of forced choice. So Sartre, with an irony that he himself could have written fiction about, ignoring his body language as body language, created one of his most famous and influential philosophical schools in the history of European thought. Yet, to do so he had to ignore his many physical maladies; he had to kill himself to achieve the fame and influence.

Now, at this point, it might do readers well to knit this analysis of Sartre's Existential nothingness into the wider conversation of our book about Nothing and nothingness. Let us do so by highlighting the vast space and conceptual contrast between Sartre's nothingness as nausea/anguish and the Nothing of the religious traditions of the world.

In several key arguments earlier, we employed the cliff-edge thought experiment to lend a visceral image to the act of standing at the end of conceptual knowing and looking over into Nothing. This act can lead to insight and to Nihilitive knowing of the divine.

Sartre also used the heuristic image of the cliff edge, yet his intent is utterly afield from ours. In *Being and Nothingness* he distinguishes between "fear" which is a fright regarding other humans while "anguish" is anxiety regarding oneself, "anguish to the extent that I am afraid not of falling over the precipice, but of throwing myself over." We envision a person timidly sidling toward the top of a cliff face and slowly coming to the realization there is no internal leash that holds them from tossing their body over the edge. This internal freedom rises up as anguish.

Anguish arises in realizing that one may make the decision to hurl oneself over the cliff unto death. Sartre expends no philosophical energy to metaphorize what is **over** the cliff edge; that realm lies outside the intent of his image. We must take for granted that, for Sartre, there is only physical death over the cliff. There is no outsight, no greeting of the divine as Nothing. Sartre's nothingness lies on the safe side of the cliff edge, whereas the Nothing of religious traditions may be encountered where safety ends.

We must recognize however that, for Sartre, the safe side of the cliff, though within his realm of philosophizing, was indeed not safe at all. For it is **on** the cliff, not over it, that the anguish of nothingness is experienced. Now, this is telling in terms of contrasting the Nothing of the world's religions with Sartre's nothingness. In the teachings of world religions, the cliff-edge represents the goal of theology, the goal of faith, and the goal of "staying awake" practices. Over the cliff lies a realm beyond the conscious and unconscious psyche; it is the realm of the super-psyche and the possible Nihilitive knowing of the divine as Nothing. Indeed, in Sartre's world there is no such realm as the super-psyche.

Again Hazel E. Barnes, the translator of Sartre's *Being and Nothingness*, is insightful regarding what she terms "mysticism," which carries essentially the same denotation and connotation as what we have termed "Nihilitive knowing." She holds that it would be quite impossible for Sartre to have composed a treatise on mysticism in that he wove all mystical elements into daily and ordinary human life. Whereas the mystic takes on a discipline toward the transcendence of self, Sartre's human being spends their life "in a futile pursuit of Being and tries in vain to escape the nothingness which it is."

For Sartre, there is not a realm beyond the cliff at all. Indeed, we must conclude that, for Sartre, there is no Nothing.

Let us make one final observation about Sartre's nothingness. Recall our statement in Section #3 discussing apophatic theology and establishing our definition regarding concepts and Nothing, "Thinking is, indeed, **about** nothingness, yet only a thinking **around** Nothing. For the purpose of building the heuristic of Nihilology, let us encode this phenomenon thus; all concepts about the nature of Nothing are actually about nothingness."

Sartre's nothingness does not lie beyond the cliff-edge of concept. He expands on its nature (being) for well over 600 pages in his magnum opus *Being and Nothingness* without implying that the nature of nothingness has not been explicated. Note that Sartre chose the abstract term *le neant*—nothingness— when he could have opted for the simple noun *le rien*—nothing. For Sartre, nothingness is within the realm of description by concept and thus falls well within our earlier definition of nothingness.

Also, recall our definitional statement in Section #4, "Nothing cannot be experienced, since by "experience" we presuppose a human being awake in the conscious psyche. The term "experience" cannot be used regarding Nihilitive knowing. (Perhaps, we could coin a new, more exact term: "inperience"!)"

Sartre's nothingness can be experienced. In fact, it is defined by being experienced as nausea and anguish. So, again, Sartre's nothingness is well-accounted for by our Nihilogical definitions.

In order to further justify and explore this interpretation, let us take a closer look at Nietzsche's *übermensch* in the context of his Nihilism, his philosophy of nothingness.

As we mentioned, the men of the Nietzsche family constituted a long line of ministers. Friedrich was very religious as a child and expected to become a leader of the flock of believers himself. The expectation that he would become a minister heralded a life-long bent to deem himself an essential persona in the history of the world. For instance, his grandiloquent philosophical novel *Thus Spoke Zarathustra* is composed in the voice of Zarathustra who was the founder of Zoroastrianism, the official religion of Persia for about 1,200 years. The form of the book mimics the canonical hymns of the religion; each parable or sermon culminates with the phrase, "Thus spoke Zarathustra." This follows in the classic Indo-Iranian tradition of poets referring to themselves, often with a self-chosen pen name, in the third person. This is also reminiscent of men of power in western political and science culture opting to refer to themselves as "he" rather than "I" as a means of emphasizing their regal quality.

But why did Nietzsche choose Zarathustra as his model? He reveals his breakthrough experience of inspiration in his book *Ecce Homo.* Just as an aside, "ecce homo" means "behold the man." These are the words proclaimed by Pontius Pilate in the New Testament of the Christian Bible as he presented the whip-scarred Jesus Christ to a mad crowd before his crucifixion. Again, Nietzsche identifies with an essential persona in world history.

In the relevant passages, Nietzsche describes walking along the coast near Genoa, Italy, "On these two paths the whole life of the first part of Zarathustra came to me, above all, Zarathustra himself, as a type: more correctly, *he overcame me...*"

Thus, Nietzsche assumed the form and function of the persona of Zarathustra (Not the Zoroastrian teachings on the oneness of God and the need for religiously based morality). He saw himself in a grandiose mirror at all times, as one yearning to be a conqueror, an argonaut of the soul of all people, a super being in the tragic play of humanity.

This is the birth of the *übermensch* or übermensch, the rather megalomaniacal persona with whom Nietzsche identifies and declares himself, in the voice of Zarathustra, to be the first among a select few who will take up the mantle of future human evolution. "Man *is* a rope, tied between beast and overman [übermensch]—a rope over an abyss... What is great in man *is* that he is a bridge and not an end: what can be loved in man is that he is an overture and a going under."

The overture opens an opera; it is a prelude to the main narrative. Human beings of nineteenth-century Europe are the prelude to the übermensch.

Recall the infamous Nietzsche declaration: *Gott is tot.* God is dead, and with this "greatest danger" out of the way, the path is open for the übermensch to become lord. As Nietzsche intones, "Dead are all gods; now we want the overman to live!"

Nietzsche, as Zarathustra, is a rather brooding, sometimes ecstatic, always puffed up prophet who finds his inspiration in the high altitudes of the mountains where he returns to rest from his prophetic labor, "I may go down again to my friends, and to my enemies too... . From the silent mountains and thunderstorms of suffering my soul rushes into the valleys."

Here two central images of Nietzsche as the übermensch conjoin. First, he is the savior on high who is the solitary source of wisdom, and that must be brought down to the masses of people below, people who, according to Nietzsche, are also low on the evolutionary scale. Zarathustra intones that the übermensch is the holder of knowledge and when he descends to "scatter his seed" it is a walk among animals.

In the final passages of *Thus Spoke Zarathustra*, Nietzsche clearly declares his usurpation of the Christian Christ. He is indeed lord, but not of heaven. "To be sure: except ye become as little children, ye shall not enter that kingdom heaven. (And Zarathustra pointed upward with his hands.) But we have no wish whatever to enter into the kingdom of heaven: we have become men—*so we want the earth*."

So, to conclude, with this analysis of the nothingness of the Jean-Paul Sartre and Friedrich Nietzsche, as well as the Nietzsche's übermensch, we are prepared to enter into Nihilology's critique of the popularizers of science. We will argue that the POPs are in fact today's torch bearers of Existentialism and the misguided incarnations of the übermensch.

(6)
Employing the Heuristic: Critique of the POPs

The Popularizers of Science and Scientism

- Brian Greene (*The Elegant Universe*)
- Alan Guth (*The Inflationary Universe*),
- Michio Kaku (*Physics of the Impossible*)
- Lawrence Krauss (*A Universe From Nothing*)
- Max Tegmark (*Our Mathematical Universe*)
- Frank Wilczek (*A Beautiful Question*)
- Neil deGrasse Tyson (*Death by Black Hole*)
- Lisa Randall (*Warped Passages*)
- Stephen Hawking (*A Brief History of Time*)

These are the popularizers of contemporary science culture: POPs as we affectionately tag them in that they are the sires of accessible wisdom regarding the nature of the universe and they function as pop stars of the intellect. They have each earned a Ph.D. degree in physics and hold positions at prestigious institutions throughout the western world. Their forefathers, much like the eons-long series of begats of the Christian Old Testament, reach back through Carl Sagan (*Cosmos*), to Richard Feynman (*Surely You're Joking Mr. Feynman*), and find their Adam in Albert Einstein (*Relativity: The Special and General Theory*).

In order to appreciate the arguments and critique that we are about to enter, it is essential to understand just how important these POPs are because of the decisive influence they have on culture and worldview of western people.

For example, Carl Sagan's book *Cosmos*, published in 1980, spent seventy weeks on the *New York Times's* bestsellers list. It was the first-ever science book to sell over 500,000 copies. five million copies have been sold internationally. It was deemed by the United States Library of Congress as one of 88 books that shaped America. *A Brief History of Time*, by Stephen Hawking, has sold approximately ten million copies since 1988. It was listed for 237 weeks as a London's *Sunday Times* bestseller, the longest residence of any book in history. For the entire summer of 2017, the best-selling book on Amazon.com was Neil de Grasse Tyson's *Astrophysics for People in a Hurry.*

These writings are immensely successful and have come to serve as the canon of science in the minds of literate people. Further, as we shall argue, they function as scripture in the modern secular world, determining the canon of reality, an essential role traditionally played by religions in both the East and the West. Now, to bring balance into this perspective and not allow ourselves to get carried away, POPs' books at this juncture do not threaten the publishing supremacy of the Bible, of which approximately 100 million copies are sold and distributed for free every year. Of course, because the Bible also serves as a sacred icon (the average household in the US owns 4.4 Bibles), it is impossible to get an accurate comparative count of actual books read. Nonetheless, it is safe to conclude that the Bible, followed closely by *Quotations from Chairman Mao Tse-Tung*, is the most widely read book over the past fifty years.

Carl Sagan's show "Cosmos" aired only 13 episodes back in 1980, yet it was a hit and made him the first scientist TV celebrity. It also functioned to showcase science on mainstream living room media. Brian Greene's internationally bestselling book *The Elegant Universe* also became the basis of a widely viewed TV show in the PBS NOVA series of 2003. No doubt his voice produced the first articulation of the words "string theory" ever heard by the public ear.

Also, with the birth of YouTube in 2005, the lectures, debates, and PowerPoint presentations of the POPs became available to the public in the visual form of which we have, of late, become most enamored. By 2019, YouTube had risen to the second most popular search engine on the Internet. And the viewership for popularizer science is quite remarkable. Lawrence Krauss's presentation on his latest book has garnered, to date, over 250,000 views. Michio Kaku's talk on his latest book has drawn over 710,000 views.

To go one step farther, readers must keep in mind that these popularizer scientists all work, write, and lecture as traditional academic scientists as well. Yet, there is one major characteristic that differentiates these two scientific milieux: mathematics.

To illustrate, Alan Guth, in the preface to his book *The Inflationary Universe,* states unashamedly that while he hopes his colleagues, who are professional scientists, would enjoy and find value in his book, it was composed for lay intellectual readers. In so doing he "avoided equations but attempted instead to explain the underlying science in words and diagrams alone..."

Mathematics, perhaps the most powerful lingua franca for all academic science communication and the foundational reason academic science remains esoteric in the minds of general literate culture, is almost universally absent from popularizer science. POPs must rely on words and images to get their views of the cosmos across to general readers. And this abandonment of mathematics has had transformative consequences: popularizer science writing has become a new genre of literature.

Elizabeth Leane, scientist and science writer, does a smart job of getting to the pith of this transformation in her book *Reading Popular Physics.* She holds it is inevitable for POPs to appropriate a wide spectrum of literary techniques. They can't, as would be normal in professional scientific discussion, assume the reader's interest and attention. They are now burdened with the novelists' task to "capture the reader's attention and imagination..."

The POPs, out of necessity for clear communication with their sought-after readership, have become storytellers. And this new cultural role is not hidden from the popularizers themselves. For instance, Lawrence Krauss's latest book is titled *The Greatest Story Ever Told... So Far.* Note here how he embraces the role of storyteller, "I am not saying in the story that I have told today that the universe came from nothing... we can give a plausible mechanism for the universe of something coming from nothing without divine intervention.... God is unnecessary or at best redundant."

Also, listen to the perspective of Brian Greene, in words that resound with poesy, as he reflects on his well-known book *The Elegant Universe.*

In the process of composing the text, he affirmed his belief that physics provides an author with the most wonderful material, that the themes of physics are the universal themes of modern humans. "We all love a good story. We all love a tantalizing mystery. We all love the underdog pressing onward against seemingly insurmountable odds..."

Amidst this multi-featured evolution in science popularization, there has grown up a new academic discipline called Science Studies. Its underlying axiom is that science has become embedded in western culture. Witness the popularity of *The Big Bang Theory* on TV with its heroization of science nerds. Science is knit in now, in much the same way that Sunday-church-going religion used to be for centuries. As we shall discuss below, this gradual knitting began to seed itself around 1550. It has taken over 400 years for the West to become a science culture.

We noted before, with a twinkle in our eye, that today's POPs hold to a line of begats that finds its Adam in Albert Einstein, and it is with Einstein that yet another feature of science popularization began to seed itself: scientism.

Einstein's general theory of relativity gave birth to a 1920s and 30s version of a media carnival whose beginning was marked by Arthur Eddington's 1919 eclipse observation which was hailed at the time to corroborate Einstein's prediction that light rays bend in a gravitational field to a much greater extent than Newton's physics allowed. Within five years after the eclipse proof, many books were written by physicists with the intent of popularizing the latest theories: relativity and quantum. Einstein's own popularization *Relativity: The Special and General Theory* led the way. Within a span of 18 months, seven editions were published.

Now, here is the point we want readers to appreciate. Many literate people of the 1920s would sense a great relief in the realization that the deep nature of the four-dimensional universe could only be fathomed by geniuses like Einstein. As Nobel Physicist Hannes Alfvén states in Claes Johnson's book, *Dr. Faustus of Physics*, "They had tried to understand science, but now it was evident that **science was something to believe in**, not something which should be understood."

And in order to more deeply appreciate how science could possibly function as an object of faith, we need to explore just briefly the history of this oddity coming to be. For 2,000 years, beginning with Aristotle, the study of nature had carried the moniker Natural Philosophy, "the love of wisdom" or the "pursuit of knowledge" about nature; nature being the manifest world in contradistinction to the realms of the divine. It wasn't until the nineteenth century that the notion of "science" came into being; until then there was no such person as a "biologist" or "physicist." The evolution from natural philosophy to science in the West is generally accepted to have its genesis in 1543 with the book of Nicolaus Copernicus titled *On the Revolutions of the Celestial Spheres.* It was indeed a revolutionary book in that it removed the earth from the center of our universe and gave birth to a movement that would remove religion from the center of the West's cultural worldview.

The early natural philosophers sought to discover the constituents and laws of the manifest world because these constituents and laws were believed to be, in themselves, manifestations of the tremendous wonder of God. In 1550, Galileo Galilei, following the lead of Copernicus, reveled, "The greatness and the glory of God shines forth marvelously in all His works, and is to be read above all in the open book of the heavens."

And then in 1596, following in the footsteps of Galileo, Johannes Kepler, upon concluding how the six planets revolve through the celestial spheres exalted in his *Mysterium Cosmographicum*, "I feel carried away and possessed by an unutterable rapture over the divine spectacle of the heavenly harmony... God himself was too kind to remain idle, and began to play the game of signatures, signing his likeness into the world..."

It is interesting to note that the British natural philosophers of this period rather called themselves "virtouosi" in that they believed they were performing a virtuous task in their explorations of the world. Robert Boyle, the famous seventeenth-century virtuoso of the Royal Society held that his work was a "religious task... the disclosure of the admirable workmanship that God displayed in the universe." And then, of course there is the most famous virtuoso of them all, Isaac Newton, who in the eighteenth century declared, "the universe bespeaks an all-powerful creator."

Thus, it is clear that these early natural philosophers sought to reveal the mysteries of manifest nature to the glorification of their God and religion. However, this glorification motivator, over time, lost its spirit, and the search became one of understanding only, not on bended knee to God. Albert Einstein may be the cusp here, as in many other aspects of science history, wherein scientists began to do their explorations without bending a knee. Ultimately this led to a scientific worldview dominating literate western culture. Science overtook religion as the bowed-to and religion become the bower. This is where we have been in the West for the past century.

Ray L. Hart, famed religion professor at Boston University, noted this state of obeisance recently in his book *God Being Nothing: Toward a Theogony*, "The cognitive traffic between theology and cosmogonic sciences is mostly one-way, moving from the sciences to theology, with theologians being the learner students."

Modern-day scientists have consciously, and with a certain amount of celebration, concluded that science has no need at all for the theology of the church in order to account for the manifest design and lawful functioning of the universe. In the POPs we even witness the happy dance and chest beating of ones who conclude that religion itself is now beside the point. This revelry gives witness to a certain wry vector in history that began with the natural philosopher Copernicus who saw himself as an apologist for church doctrine and culminated with the POPS who see themselves, not only as victors in the war for supreme explainers, but also as antagonists to church doctrine.

However, the POPs are not only antagonists to the church, they are also inheritors of the church. For indeed, with the embeddedness of science as the purveyor of our modern literate worldview, many of the functions of the church have now migrated to the bailiwick of the POPs. The POPs themselves are unaware of, and would indeed disavow, any obligations to their believing public which are *de facto* inherent to their offices as elders in the scientific worldview. Nonetheless, the facts of the matter are clear for all those of us who have eyes to see. Recall our discussion before regarding Hannes Alfvén who stated that, for those of us who are stymied by the mathematics of four dimensions, ever since Einstein science has served as a comforting belief system whose stories provide us with a vision of the manifest universe to believe in. Alas, science has become its own "ism."

A perfect expression of faith in science is heard in the popular book by California Institute of Technology cosmologist Sean Carroll: *The Big Picture: On the Origins of Life, Meaning and the Universe Itself.* Carroll states that we as a culture are coming to "recognize that the universe is not buoyed up by a transcendent being" nor by transcendent thinking. He then comments on his own assertion, "The meaning we find in life is not transcendent, but it's no less meaningful for that." (This comment foreshadows our critique below regarding POPs as Existentialist philosophers.)

Carroll calls forth a wonderful and evocative image from the old Looney Tunes character Wile E. Coyote who, while chasing the illusive Road Runner, inevitably goes headlong over a cliff. Wile E. then stops in mid-air, doing a wide-eyed mugging into the camera, as it begins to dawn on him what has happened. "Wile E. Coyote would have been fine if he had been equipped with one of those ACME-brand jet packs, so he could fly around under his own volition. It's time to get to work on our own conceptual jet packs."

Carroll states his belief that all conceptual ground has disappeared under our feet and we are just beginning to work up the courage to look beneath us. Carroll's position, in simple terms, is that faith in religious myths and explanations used to buoy us up, keep us levitating, but now religious myths and explanations have been dismissed by western science culture. Yet, there is still a moment or two in time, while Wile E. is mugging the camera, for science to come up with myths and explanations, "conceptual jet packs," to serve the public's need for belief, and keep Wile E. from plummeting to the canyon floor.

Now, employing our heuristic of Nihilology, readers are aided in analyzing Carroll's position and discovering that there are two basic flaws therein, two flaws common to many POPs cosmologists. Before elucidating these flaws, recall from our earlier discussions that Nothing lies over the cliff edge of rationality. Nothing lies over the cliff edge of conscious thought and emotion and over the cliff edge of unconscious forces and parallel processing, the super-psyche wherein the divine as Nothing can be met by the person as Nothing. This is a meeting for which we coined the term "inperience."

So first, Carroll takes advantage of a metaphoric image that is *a priori* grounded in a worldview where there exists an object of faith that transcends reason, a realm over the cliff edge. This transcendent realm of faith is where the image finds its explanatory force, its heuristic value. Yet, according to Carroll, and to all other POPs as well, there is no realm beyond that of reason. Wile E. Coyote does not need to be buoyed up above the abyss of the canyon for, indeed, in Carroll's world there is no abyss, no canyon to levitate above.

This misappropriation of a heuristic image from a religious worldview of transcendence is one of the negative consequences of POPs evolving into storytellers. Deprived of their familiar language of mathematics, they are still quite immature in their facility to appreciate the logic of story images. Thus, they tend to borrow images from literature that actually work against the very argument they are putting forth!

Second, recall once again the quote by Hannes Alfvén, "[Laypeople] tried to understand science, but now it was evident that science was something to believe in, not something which should be understood." Popularizers of science are attempting to co-opt the realm that people of western culture have, for centuries, reserved in their worldviews for belief in a transcendent order. **Nothing** lies in the realm of the transcendent. **Nothingness** lies in the realm of reason, and this is a distinction that needs to be made. For it clearly reveals to laypeople the insufficiency, and perhaps even the contradiction, of having faith in natural world things, having faith in nothingness.

Remember, we established earlier that Nothing lies outside the realm of rationality. Nothing lies over the cliff edge of all continua of rational knowledge and of experience itself. The search of the POPs is by their own definition within the realm of rationality. Faith is misplaced and detrimental here. Misplaced because the realms of rationality can be known via empirical means yet will never be known if we make it an object of belief. Nonetheless, POPs encourage this sort of belief in their work.

As we shall see below, the POPs are themselves blindly faithful regarding beauty and unification. Faith in their science further translates into faith in themselves which inflates them into the role of übermenschs in culture. They admonish laypeople to hold no stock in faith yet reap the rewards of those laypeople who do have faith... in their stories and in them. Further, mature faith always serves as a heuristic, goading one on to understanding. Faith in science, as described, serves as an anti-heuristic; it hinders the seeking of understanding.

So these are the central issues we will now address, making way for our bold assertion that it is rational for popularizes of science to accept and learn from the findings of Nihilology in general, and the differentiation of Nothing and nothingness in particular. First, the books of popularizer scientists are being published in voluminous quantities and are widely disseminated among literate society in the West. Second, POPs are becoming popular media stars, TV personalities, and newsworthy characters. The public persona of the POPs has become, as we shall argue, an active goad and guide of scientists which in turn becomes a distorting goad in their science theorizing and empirical research.

Third, the general culture is becoming aware of science as an acceptable and dominating worldview, that is, science is embedded therein. Fourth, POPs are resorting to the narrative techniques of literature in the form of storytelling. They are making myths for the modern West; this is just one among many ways that POPs are functioning in our culture that are not appreciated by the literate public and, deviously so, often not acknowledged by the POPs themselves. Finally, readers of popularizer writing are attracted to it in a large part by the function it serves as an object of faith, something to believe in.

Modern Science Usurping the Role of Religion in Western Culture

- *Knocking on Heaven's Door: How Physics and Scientific Thinking Illuminate the Universe and the Modern World* (Lisa Randall)
- *The Varieties of Scientific Experience: A Personal View of the Search for God* (Carl Sagan)
- *Visions: How Science Will Revolutionize the 21st Century* (Michio Kaku)
- *The Greatest Story Ever Told—So Far* (Lawrence Krauss)
- *The Lightness of Being* (Frank Wilczek)

As popularizer scientists have become embedded as the ultimate explainers of the creation and workings of the universe their stories have also evolved into the scriptures of western culture. One way that POPs usurp the former role of religious scripture is by making use of the evocative power of religious book titles to support their own scientific worldview. POPs, of course, deny that they benefit from religious titles and references, yet it is clear for all who read their books that this appropriation of evocation is fully at play.

For example, Lisa Randall, in her *New York Times* bestseller *Knocking on Heaven's Door* makes reference to Bob Dylan's 1975 song. The theme of the song is one of darkness at death before entering heaven:

Mama, put my guns in the ground
I can't shoot them anymore.
That long black cloud is comin' down
I feel I'm knockin' on heaven's door

Randall is stating that she, along with other popularizers of science, will throw light on conundrums that have not been solved by religious experience but will be solved by science.

Another example, *The Varieties of Scientific Experience: A Personal View of the Search for God* by Carl Sagan recalls William James's *The Varieties of Religious Experience,* a book comprised of James's Gifford Lectures on the nature of religion delivered back in 1901-02. In so doing, Sagan implies that scientific knowing is an evolved from beyond religious knowing. Further, if read critically, the title is misleading in that Sagan is drawing readers with a magnet of religion, yet his search for God is ingenuine. It was a foregone conclusion on the author's part, before a word of text was composed, that his book would deny the existence of God.

Michio Kaku joins in the writing of science scripture with his book *Visions: How Science will Revolutionize the 21st Century.* Kaku makes use of the religious connotations of the term "visions," bringing to the reader's mind's eye the images of saints and mystics encountering the sacred. Then he boldly states, regarding the subtitle, that scientists will, in the near future, usurp the function of gods as creator of nature's form, "[Physicists] are on the cusp of an epoch-making transition from being *passive observers of Nature to being active choreographers of Nature.*"

It is quite revealing to note that, in his writings, Kaku capitalizes the term "Nature," thereby personalizing and sanctifying the notion, while he calls on the lower case when discussing the functions of god(s). For example, in an article titled, "Shocker: The Universe was Created," he tells of the powers of the Greek gods to fly in a chariot (Apollo) or manifest things through thought (Zeus) then states, "By the year 2100, we will have the power of the gods."

Kaku is not shy in prognosticating the upcoming sanctification of science. He continues here, misquoting Arthur C. Clark, "Any sufficiently advanced technology is indistinguishable from divinity." Kaku outdoes Clark in his foretelling here in that Clark actually said "magic," not "divinity."

Now the master among the POPs of usurping religious titles to empower their scriptures of the scientific worldview is Lawrence Krauss. In his most recent book, *The Greatest Story Ever Told—So Far*, he is referencing the Max Von Sydow movie of 1965 with the same title that recounts Jesus's life on the big screen; it also harkens to a book and radio series of the 1940s about the four gospels of the Christian New Testament. With this title, Krauss boldly intones to his readers that science, as witnessed in his scriptural text, is taking over the role of the gospel tales of Jesus. His book is the POPs version of the good news.

Also, in a YouTube video based on this book title "Hidden Realities: The Greatest Story Ever Told... So Far," Krauss states the following, "In my opinion, the Large Hadron Collider is humanity at its best. It is the Gothic cathedral of the twenty-first century." With this statement, can there be any doubt that Pops see their science as the usurpation of religion?

Finally, let us look at one more telling example of this phenomenon of POPs using religious evocation in the titles of their science scriptures. One of Frank Wilczek's earlier books is called *The Lightness of Being*. *The Unbearable Lightness of Being* is the title of a famous novel from the 80s by Milan Kundera. It takes place during the era of the Prague Spring and reveals people's struggle to find pattern and meaning in the seemingly random, strange, and sometimes cruel world. Wilczek reflects on his own book title, "through science... coming to understand the deep structure of reality has helped to make Being seem not merely bearable, but enchanted—and enchanting. Hence 'The ~~Unbearable~~ Lightness of Being.'"

So, from Wilczek's perspective, being alive in the universe is unbearable without science, yet with science living becomes enchanted and reveals its meaning. This is, of course, a function that has traditionally been provided by religions to beings alive in the world.

Now, as we look inside the books of these leaders of popularizing science, we are struck by the fact that the *a priori* personal religious positions of the authors are often revealed in the prefaces. Before these physicists begin to tell of their storied lineage of Copernicus to Newton to Einstein, they unwittingly let readers see into their worldviews. Wilczek's *A Beautiful Question* is the epitome that first brought this notion to our notice.

On the cover of the book there appears a beautiful, author-commissioned drawing of the Great Symbol of the Chinese religion, Taoism, often referred to as the yin and yang sign. The sign resembles two carp in a playful circle that recalls the myth of the waterfall called Dragon's Gate. It is said that carp often try to jump up the falls and if they succeed, they are magically transformed into auspicious dragons. Wilczek gives us his take on this myth. "With a sense of humor, we may associate this event with the transformation of virtual particles into real particles... Alternatively we may identify ourselves with the carp and their strivings with our quest for understanding."

This display of the Taoist Great Symbol in conjunction with Wilczek's words give witness to three important points that Nihilology makes in critique of the POPs. First, we see here, as was the case with book titles, a usurpation of a religious symbol, transforming the carp of the Dragon's Gate waterfall into a heuristic image for science. Second, when Wilczek implores us to "identify ourselves with the carp and their strivings with our quest for understanding," a quest that inherently finds its goal when we "transform into auspicious dragons," he witnesses an important contemporary feature of popularizer science, the re-introduction of the personal self into the empiricism of the scientific endeavor. We will discuss this re-introduction in depth in the next theme.

And thirdly, when Wilczek, along with many of his popularizer colleagues, invites readers to join in a "quest for understanding" the deep structure of reality, it is impossible to distinguish their voices from those of religious evangelists or mystics. Our sense is that they both sound the same in ringing out a rallying cry for all to plummet the mysteries of the world beyond common experience.

And here is another call-to-join, intoned by Max Tegmark, professor of physics at MIT and co-founder of the Future of Life Institute, from the preface of his most popular book *Our Mathematical Universe: My Quest for the Ultimate Nature of the Universe.* "My goal... is to invite you along on my personal journey of exploration... this book is scientific autobiography of sorts... it's my personal quest for the ultimate nature of reality... "

We are not the first to recognize the likeness of these calls-to-join by popularizer scientists and religious calls-to-join. Lawrence Shehan, professor of Education at Mills College in New York, conducted an ingenious study of the worldviews of scientists and religious mystics.

The study consisted of presenting readers with pairs of gravitas-bearing statements regarding the nature of the universe and human life therein. They were then asked to determine if the authors of the statements were scientists or religious mystics. Shehan found that his subjects, in statistically significant numbers, could not tell from the statements who was who. This finding supports our critique that popularizer scientists are usurping the roles of religion in our scientific culture of the West. In this case, they have taken over the role of evangelists in their call to join the scientific quest.

Now let us go yet another step. Once we begin to read the actual text of these bestselling books of the POPs, we note immediately that these authors are motivated in their scientific explorations by the ambient urge to find oneness at the core of the manifest cosmos. This urge reveals itself in two intertwined themes: cosmic essence and unified theory, both of which are expressions of traditional themes and goals of religious thinkers and their sacred texts, wherein cosmic essence is termed "monism" and unified theory is termed "monotheism."

Perhaps the most abstract universal essence told in the popular science canon is mathematics, purported to have been discovered by Max Tegmark. And oddly enough, in his book *Our Mathematical Universe: My Quest for the Ultimate Nature of Reality*, Tegmark has elected to convince his readers, in keeping with the essential *a priori* of POPs literature, without the benefit of mathematics. He argues that a unified description of the universe must have no smack of the human mind about it, thus it must be free of concepts. A theory of everything must consist of equations alone. Then comes Tegmark's most pithy statement, "Our external physical reality is a mathematical structure... That is, we don't invent mathematical structures—we discover them, and invent only the notation for describing them."

It is important to re-emphasize Tegmark's point here, as it is a canon-wide position averred by all POPs regardless of the specific essence being put forth. The essence of the manifest universe, in Tegmark's view, **is** mathematics. Mathematical formulas are merely a means of describing this essence. The essence itself is out there, objectively existent. Mathematics is not a theory or other human created proposition; mathematics is "discovered."

Now, perhaps the most aesthetic cosmic essence among the POPs canon is suggested by Frank Wilczek in his book *A Beautiful Question: Finding Nature's Deep Design.* He perceives a world wherein the mind of the human being is by nature one of "miraculous harmony" with the material world. This harmony finds its roots in beauty. "When we form expectations that turn out to be correct, we experience pleasure and satisfaction. Those reward mechanisms... stimulate—indeed, at base they *are*—our sense of beauty."

Brian Greene focuses in on a more reductionist essence. "According to string theory, the universe is made up of tiny strings whose resonant patterns of vibration are the microscopic origin of particle masses and force charges."

In order to be true to trending POPs literature, we must mention Ray Kurzweil's prophecy regarding cosmic essence. Kurzweil is a computer scientist and futurist. Though neither physicist nor cosmologist, he is one of the world's most well-known POPs. Author of nine books, seven of which ascended to the *New York Times* bestseller list, he is quite certainly the most influential futurist in the world.

In his book *The Age of Spiritual Machines*, Kurzweil foretells of intelligence as an emergent cosmic essence. This is indeed the most New Age of the purported essences in that it does not exist in the present but will come to be in a future phase of manifestation called Epoch 6. Note this fantasized discussion with Bill Gates:

(Kurzweil) "We need a new religion...."

(Gates) "So, is there a God in this new religion?"

(Kurzweil) "Not yet, but there will be.
Once we saturate the matter and energy in the universe with intelligence
it will "wake up," be conscious, and sublimely intelligent.
That's about as close to God as I can imagine."

We must insert here that Kurzweil is quite irresponsible regarding terminology in his discussion of essences. At various moments, he equates or substitutes in usage the terms "intelligence" "consciousness," "spirituality," and even "empathetic perceptions." He indeed prophesies an emergent essence to the cosmos, yet his loose language does not allow us to conclude precisely what that essence will be.

Finally, let us touch upon the writing of David Bohm regarding cosmic essences. Bohm is often called one of the most significant and unorthodox theoretical physicists of the last century. As witnessed in his most influential book *Wholeness and the Implicate* Order, the so-called "holomovement" is the most mystical essence to be propounded in the POPs canon. Indeed, it gives scientific voice to a classic monism of esoteric religion, yet it is merely an echo of the mystical in that it refers only to manifested reality. "We have seen that in the 'quantum' context, the order in every immediately perceptible aspect of the world is to be regarded as coming out of a more comprehensive implicate order in which all aspects ultimately merge in the undefinable and immeasurable holomovement..."

Each of these cosmic essences—mathematics, beauty, vibrating strings, intelligence, holomovement— is put forth by a popularizer of science to serve a purpose: to aid lay readers in understanding the manifest reality. Each POP, with their particular essence, offers lay readers what might be called a pathway to understand the objective world. Remember, POPs hold that these essences of the manifest are objectively existent; they are out there in the manifest to be discovered.

Yet, from the overarching perspective of Nihilology, gleaned from the pattern of teachings witnessed in science, religion, and philosophy, there is no such essence to the objective manifest world absent the subsuming presence of Nothing. POPs search for one objective essence, the scientific reflection of the religious search of monism, will never be complete if it does not welcome nothingness **and** Nothing into its "one-ism."

Recall our analysis in Section #3 where we elaborated on the religious-philosophical notions of the East and West regarding the interrelationship of Nothing and nothingness. There we recognized that any manifest object must be perceived "surrounded" by Nothing. One might say that things float in Nothing as their context, and this context lends reality to the thing.

In that same Section #3, we concluded that the objective manifest world is not in any way separate from Nothing. Both manifestation and Nothing are present at all times, even during manifestation. Thus, manifestation, in and of itself, is not actually a reality at all and thus cannot have an essence.

So now, let us turn to the twin theme of this ambient urge on the part of popularizers of science to find oneness at the core of the manifest cosmos: unified theory. This search exhibits the POPs primal drive to function as the supreme explainers of objective reality. As such, they often employ high-flown terms for their goal such as "grand unified theory" or "theory of everything."

Witness the perspective of Alan Guth, professor of physics and cosmology at the Massachusetts Institute of Technology. In his popular book *The Inflationary Universe: Quest for a New Theory of Cosmic Origins,* Guth quotes Einstein on his motivation toward the discovery of unitive physical laws. He then soars regarding the motivation of today's POPs to find a single theory of everything. He holds that integration of theories has inspired the greatest successes in the history of science, noting the classic theory of the atom by Democritus and the law of universal gravitation as articulated by Newton. "Many physicists, including the author, hope that this list of momentous accomplishments will someday include a grand unified theory."

Also witness this statement by Brian Greene from his book *The Elegant Universe: Superstrings, Hidden Dimensions, and the Quest for the Ultimate Theory.* "Einstein was simply ahead of his time. More than half a century later, his dream of a unified theory has become the Holy Grail of modern physics."

Note how, for Greene, a quest for a single cohesive explanation of the constituents and laws of the objective world has taken on a mythic air, with popularizer scientists playing the heroic roles of Arthurian knights in search of the chalice of eternal youth... also known as the Cup of Jesus.

Now this scientific quest for a single unified theory of the universe has its parallel in the theological search for belief in one god, or monotheism. This parallelism is not an esoteric finding on our part; it has been noticed and documented by many writers. A perfect example comes from Marcelo Gleiser, the Brazilian physicist resident at Dartmouth, who we called on earlier for helpful analysis. He notes that as far back as the birth of modern science in the early 1600s, we can witness a motivating passion, not unlike religious fervor, that the universal puzzle can be solved. "British Physicist Stephen Hawking... compared the achievement to 'knowing the mind of God'... Is this Final Truth simply the scientific incarnation of the monotheistic tradition of the West, a yearning for a God that reason exorcised from spiritual life?"

As we outlined earlier, natural theologians, beginning—in the western lineage at least—with Copernicus and Galileo, researched the constituents and laws of objective reality with faith that their discoveries would glorify the wondrous mystery of God's creative power. With the era of Einstein, glorifying God lost its motivating strength, and the search for discovery and empirical understanding became its own motivator.

Nihilology makes us aware of two points of critique here. One, the momentum toward glorifying the divine on the part of natural theologians, in an intellectual expression of Newton's first law, has been conserved by leaping to the self-glorification of the POPs themselves as übermenschs. We will discuss this dynamic in depth below. Second, the quest for faith in a divine maker was not at all exorcised by rationality, regardless of the protestations by popularizer scientists. It has resurfaced, and this is obvious for all who reflect on the history of science, in the guise of faith in the discovery of a grand unified theory.

Now, what is the end goal of the grand unified theory? What will the effect of discovering the Theory of Everything be? Here is Brian Greene again, giving voice to a typical POPs answer to this question, "The ultimate [unified] theory would provide an unshakable pillar of coherence forever assuring us that the universe is a comprehensible place."

So this is the goal: to find a comprehensible universe. It's true that we can think of the universe and have assurance of our thoughts. However, from the viewpoint of Nihilology, this assurance is a false goal. As we have stated before, Nothing is eternally present, not lost at the creation of the manifest world. It is that subsuming presence which makes the manifest real and functional. Thus, to say that the grand unified theory will make the world comprehensible is to say that Nothing is comprehensible, and we have learned that Nothing is by its very nature beyond rational comprehension. It lies beyond the cliff edge of reason, affirmable by faith, and inperienced as a meeting of Nothing and Nothing. Thus, there must always remain something elusive, beyond rational comprehension about the objective world, something that POPs should be satisfied and fulfilled at not being able to comprehend.

Also, earlier, we recognized that the most fruitful way to think about Nothing is the *via negativa,* the way of negation, also called apophatic theology or philosophy. The urdoctrine of the negative way is that it is not possible to make a true statement that encompasses the nature of Nothing. Indeed, any conceptualization about the nature of Nothing alters its nature in the same way that the very act of watching the behavior of electrons alters the nature of the electron itself. It is no longer Nothing about which we are thinking; all concepts about the nature of Nothing are actually about nothingness. Later, we will suggest it is a rational strategy for popularizers of science to accept the input of Nihilology toward the improvement of their theorizing and empirical research. This will create a new evolved form for the POPs that we call apophatic science.

In these two quests, the search for cosmic essence and the search for the grand unified theory, popularizer science has usurped, and unknowingly inherited, two classic seekings of religion. These quests parallel those of religious monism on the one hand and religious monotheism on the other (one) hand. Regarding both quests the POPs can learn and gain clarity from Nihilology, without which the search will end in failure.

And, from the strength of our arguments in this section thus far, popularizer scientists, were they to follow their own dictates of scientific argument, must accept the obvious. POPs are usurping the roles of religion in western culture, while denying they have inherited any of the obligations of people's faith in their science and in them. They are blind to the fact of their own self-created scientism.

Science Transition into Religion Lite

Earlier on, we related two experiential stories that led us to outsights regarding Nothing. The first story revolved around an essay Gustavo wrote in graduate school critiquing eighteenth-century Scottish philosopher David Hume on his teachings of causation from Nothing. It led us to our understanding that Nothing cannot cause something to come to be; only nothingness can accomplish that. The second story was of Gustavo's professor's projected image of the Rutherford gold foil experiment that illustrated the vast region of the atom is essentially empty. The image stimulated an essential outsight as to the nature of Nothing. Now, there is one more story we would like to tell, this one about how we came to a realization regarding the essential role of faith as a guide in the investigations of popularizer scientists.

In the summer of 2008, Gustavo was walking along a wide lakeside beach in Algonquin National Park. He happened upon a plaid pattern of squares carved in the damp sand just above the wave line. It was eight squares by eight squares with alternating squares etched by small troughs as if dug out by its maker with a stick. An assortment of figurines was placed about the grid. He looked curiously at these pieces: their craftsmanship, the fact they come in two colors, three of the shapes have two of each color, two separate shapes have only one of each color, and one rather basic shape has eight.

This forsaken array stimulated a reverie for him. Being in a playful beach-day mood, he began to create a set of proto-rules for a match game using the pieces. As he encountered problems—uninspired consequences—with his initial set of rules, he amended them and further amended certain amendments as he narrowed in on a workable game paradigm.

Then it occurred to Gustavo, another person may come walking the beach, look at the pieces on the sand checkerboard and make up a solitaire game. Still another beachcomber might entertain themselves by simply making arrangements with the pieces. Yet, he reminded himself that the smarter and more imaginative he was the better his game could be, which could objectively translate to success in convincing other beach walkers to suspend their drive to create for themselves and play according to what would no longer be his rules, but their rules.

Now nobody within the beach world of Gustavo's reverie knew that this checkerboard and pieces would constitute in the daily world a classic board game called chess and that for chess there is an absolute set of rules that yield the best possible game one can play. And, within his reverie, no beachcomber could prove rationally or mathematically that there exists a best possible game or even that the best possible use of the sand checkerboard is a game at all. Nonetheless, as he imagined a whole array of rules and strategies for improving match play on his sand checkerboard, an intuition nagged at him that there must exist a best possible game.

Chess is the best possible game one could create given what was laid out on the beach. Surely better games can be created with fewer pieces or by increasing the complexity of the board-space, but considering that different pieces should correspond to different functions all working together to achieve a desired outcome, playing on an 8x8 checkerboard, no player can do better than chess. This intuition brought up in him yet another: the best possible game cannot be a creation of his own thoughts. The best possible game could not be cracked by him from within his own head. It must exist in the objective reverie beach world.

At that moment, Gustavo exited his reverie, bearing fruits that are directly applicable to our critique of popularizer scientists for, indeed, POPS are seeking the "Best Possible Game" for the checkerboard of the manifest universe. This best possible game takes the form of the grand unified theory.

We pointed out before how faith in one god—monotheism—supposedly put to rest by the rationality of science, has been reborn among the POPs in the guise of faith in a unified theory of the universe. This rebirth has been attested to repeatedly in the writings of POPs and those who study the history of science. Gleiser said the grand unified theory is "simply the scientific incarnation of the monotheistic tradition in the West." Hawking coined it "knowing the mind of God." Guth stated, "Indeed unification has been the inspiration for many of the greatest achievements in the history of science." According to Greene "a unified theory has become the Holy Grail of modern physics," and it will "forever assure us that the universe is a comprehensible place."

Our outsight revealed that POPs are moved to discover the grand unified theory out of a faith that the universe is a reality that indeed **has** an objective best-possible-game nature. The universe is such that a grand unified theory is inherent to its nature. Just as was the case with universal essence, POPs hold that the grand unification is objectively existent; it is out there in the manifest to be discovered. Just as Tegmark exclaimed, "The essence of the manifest universe **is** mathematics." So too, the nature of that universe **is** unified and can be described in a theory as such.

And finally, Gustavo's reverie on the beach checkerboard convinced him that belief acts as a heuristic for intellectual understanding. In this case, the POPs' faith in the unified nature of the universe, and in the power of a grand unified theory to embody its nature, work as a tool for deeper understanding and expression of the universe as single.

Even though the popularizers of science, to a man or woman, deny that belief is functioning in their empirical endeavors, we can employ the fruits of Gustavo's thinking to argue quite clearly that their denial ignores the evidence at hand, as witnessed in the writings of the POPs themselves.

We have already discussed, from a myriad of angles, just how faith serves as a trans-heuristic toward outsight and a heuristic toward intellectual understanding. We relied on theologians and psychologists to make our case. Remember Anselm of Cantebury's profound declaration in his Proslogion, "For I do not seek to understand in order to believe, but I believe in order to understand. For I believe even this: that unless I believe, I shall not understand." We also learned from Gordon Allport, who in his book *The Individual and His Religion* held that mature faith impels believers to discover new understanding beyond the bounds of current belief, that "Feeling that we already have the one final and complete truth would not be heuristic and therefore would not be mature."

I.J.R. Aitchison, the famed theoretical physicist from Oxford University and leading voice in gauge theory, in his article "The Vacuum and Unification," echoes and bolsters our perspective. He points out that the known laws of physics, especially over the last century, have evolved toward ever greater unification. "This drive toward unification has... acted as a powerful heuristic device in constructing phenomenologically successful theories..."

There is one more belief among popularizers of science that we should add here. POPs have great faith in their own rational powers; they believe the grand unified theory will be discovered via empirical and rational thought... **alone** without the aid of the divine or any other agent. This belief is attested to throughout the POPs canon: Lawrence Krauss, Alan Guth, Steven Hawking, they all proclaim it. Lisa Randall rides the crest of faith in the rational. She opines that much argument arises from primal beliefs as to whether humans have the tools to find the truth of manifest life on their own. Religionists say no. Randall disagrees, "The universe is humbling. Nature hides many of its most interesting mysteries. Yet, scientists are arrogant enough to believe we can solve them."

So, POPs make use of the heuristic power of faith already; they don't need to be convinced through our critique to do so. In the final section of our book we will argue that popularizers of science would evolve in their theory and practice were they to make the rational decision to believe in Nothing. Yet, for now, let it suffice to conclude that POPs depend on and are buoyed up by (1) their faith that a unified theory, a monotheory, does indeed exist, (2) their faith that the manifest universe is indeed a law-unified reality, and (3) their faith that they have the mental capacity to discover this unity. All three of these are beliefs, unproven and without empirical basis. POPs indeed have faith, and it aids them in becoming better scientists!

Let us go one step farther in this critique of popularizer scientists as faithful. Indeed, faith in the goal of a grand unified theory is clear, yet it is also faith that **guides** the POPs quest for unification. As we read through the entire canon of popularizer scientists, it came as a jolt to our *a priori* sense of these leaders of the empirical that there is a widespread, and proudly proclaimed, belief in the guiding power of beauty!

The leading voice among POPs in preaching for beauty is Frank Wilczek as witnessed in his popular book introduced earlier, *A Beautiful Question: Finding Nature's Deep Design*. Therein he states that trust in beauty has often led to profound theoretical payoff throughout the history of science. He also looks forward and avers that supersymmetry, a step toward the grand unified theory of all reality, will be found through the guidance of beauty. "In beauty we trust, when making our theories... truth is highly desirable, but it is not the only, or even the most important, criterion..."

We hear Wilczek fully validating the heuristic nature of faith. His faith is in beauty. Yet, faith in beauty is not only his personal guide, according to Wilczek it has also guided the entire field of physics. He points to James Clerk Maxwell, the mid-nineteenth century unifier of electricity and magnetism, as the first among a long line of believers in beauty. "Logical consistency is required, but hardly sufficient. Rather it was beauty and symmetry that guided Maxwell and his followers—that is, all modern physicists—closer to truth..."

This affirmation of faith should remind readers of our earlier discussion of Watson and Crick being guided by beauty in their discovery of the molecular form of DNA. Their double helix was experienced as an aesthetic creature, not a rational discovery. As Crick exclaimed, "It's so beautiful you see, so beautiful."

Now, though Wilczek is the most extroverted about his faith in beauty, this belief is pervasive in the canon of popularizers of science. Listen here to Richard Feynman's avowal of the aesthetic motivation for René Descartes' famous work on the refraction of light, "And what do you think was the salient feature of the rainbow that inspired Descartes' mathematical analysis?... I would say his inspiration was that he thought rainbows were beautiful."

Feynman himself was also inspired by his foundational belief that the manifest world is profoundly beautiful. He marveled at the fact that inanimate and living things appear so differently and act so differently yet are organized by a unified set of laws. He called it the "mathematical beauty of nature."

And listen here to Physicist Andrew Strominger, Director of Harvard's Center for the Fundamental Laws of Nature, as he intones his belief that if a theory is beautiful it is held to be akin to the inherent structure of the world. He refers to Superstring Theory, another of the steps along the path to the grand unified theory of the manifest universe, and says that there is not a single scientist who is smart enough to invent the theory, so it must be that the theory has been discovered as an objective existent. In words that echo Max Tegmark he says, "It's beyond our imagination to invent such a beautiful and powerful mathematical structure."

And finally, let us note the words of Steven Weinberg, Noble Prize winner in Physics for his work toward unifying two fundamental forces of nature: the weak force and electromagnetism. In the book *Dreams of a Final Theory: The Scientific Search for the Fundamental Laws of Nature* he brims with hope that science will soon reveal the ultimate and final laws of the universe. "Time and again physicists have been guided by their sense of beauty not only in developing new theories but even in judging the validity of physical theories once they are developed. It seems that we are learning how to anticipate the beauty of nature at its most fundamental level."

So faith in beauty serves popularizer scientists as a heuristic device toward the discovery of a unified universe. In the concluding chapter of our book, we will recommend the POPs consider taking on a faith in Nothing as another heuristic toward evolving their understanding of the world. Yet, to culminate this present phase of our critique, please note that not a single POP gives forth an analysis or even a reflection on the experiential basis of beauty. Do they hold in common an argued and agreed-upon definition of beauty? How do these men and woman know they are in the presence of beauty? What is the experience of beauty? Not one examination of the aesthetic experience is found in the canon, as far as our research has shown, and we have been paying attention to see if anyone approached the issue. Faith in beauty is the guide that will lead to the grand unified theory, yet there seems to be no POP who is reflectively feeling the mood of beauty.

However, this shying away from experience is not the case regarding night sky gazing, an encounter that essentially all popularizer cosmologists and astrophysicists point to as a source of inspiration to pursue their field of study or the fertile ground from which new thinking blooms. The experiential aspect of looking up at the night sky is commonly referred to by the POPs as awe or the sublime. It is this experience of awe that we had in mind when we titled this theme "Science Transition to Religion Lite." Indeed, for at least 2,000 years awe had been regarded as a religious experience. Then in the late eighteenth century the philosopher Edmund Burke, in his book *A Philosophical Enquiry About Our Ideas of the Origins of the Sublime and Beauty,* perpetrated something quite radical in the culture. He secularized awe! So, even though popularizers of science still today experience awe, it has been stripped of the divine. Thus we say they are having "lite" religious experience.

Just as a side note, this secularization of awe, reinterpreting the experience to rid it of the divine, is occurring concurrently with the transition from Natural Theology, as we discussed before, to what we now know as science. Scientists took the divine designer out of nature at the same time that they were taking the divine out of awe, thus secularizing nature and secularizing one of the pinnacles of human experience.

Let us foreshadow one intention for this critique of awe among popularizers of science. In earlier parts of our book we put forth Nihilology's unique model of outsight; now we will expand on that model to include the experience of awe as well. (Such a model is, of course, missing in the POPs canon.)

Now, beginning as far back as the 19th century, popularizers of science have proclaimed the experience of awe to be a hyper and intense emotional state that serves to drive scientific achievements. And, again as we have seen often, Albert Einstein was the first to be heard by the public in this proclamation. "I don't try to imagine a personal God; it suffices to stand in awe at the structure of the world, insofar as it allows our inadequate senses to appreciate it." He also proclaimed, "I maintain that the cosmic religious feeling is the strongest and noblest motive for scientific research."

Carl Sagan stands firmly in the lineage of Einstein, even borrowing much of the Einsteinian language regarding awe and science, "By far the best way I know to engage the religious sensibility, the sense of awe, is to look up on a clear night... Hidden within every astronomical investigation, sometimes so deeply buried that the researcher himself is unaware of its presence, lies a kernel of awe."

It is essential to note that Sagan, and in this he represents the common view of all POPs, differentiates between religious sensibility and belief in a divine being. Thus, POPs acknowledge the experience of awe as necessary and effective in giving birth to careers and new scientific theory, but they have no sense of personal god or belief in doctrine associated with a personal god.

But, note these words of Sagan as well, "We are, in the most profound sense, children of the Cosmos... Our ancestors worshiped the Sun, and they were far from foolish... If we must worship a power greater than ourselves, does it not make sense to revere the Sun and stars?" Sagan reveals his ambivalence of belief by capitalizing "Cosmos" and "Sun," making them into proper names, personalizing them, revealing his kinship with the reverent ancestors from whom he seemingly tries to distance himself.

Richard Dawkins, emeritus professor of evolutionary biology at Oxford and world-famous POP as leader of the four horsemen of new atheism, is also in the lineage of Einstein. In his book *The God Delusion*, which has sold upwards of four million copies, he relates a story of a youngster gazing at the stars of an African night sky and becoming bedazzled at the sight of Orion, Cassiopeia, and Ursa Major. His eyes become full of tears "with the unheard music of the Milky Way." Dawkins then says that this boy could have been him having "a quasi-mystical response to nature and the universe" as is "common among scientists and rationalists. It has no connection with supernatural belief."

Again and again, POPs declare the sentiment of Einstein as representing their form of non-religion. Part of this declaration is to rigorously hold that the experience of awe they feel in looking up into the stars at night must be differentiated from any sort of religious belief. Yet, note here, as was the case with Sagan, that Dawkins is ambivalent in his declaration. He calls awe "a quasi-mystical response." Now this may mean that there is such a thing as mystical experience but his wasn't quite it, or perhaps, his resembles what people call a mystical experience. But there is no such thing in reality, and thus his wasn't one.

Let us quote two more popularizers of science regarding the role of awe in science, namely Oxford mathematical physicist Roger Penrose and Harvard's theoretical physicist Lisa Randall. These two often leap into refreshing lyrical flights in their descriptions of awe and their words here, presented in stanzas to emphasize their poetry, opportunely support our interpretation of awe at night skygazing and our model of awe that we will present as the culmination of this discussion.

Penrose, whose work in part attempts to accomplish a step toward a grand unified theory, tells a story of awe in his *Sunday Times* top 10 bestselling book *The Road To Reality*:

Antea, a postdoctoral student of physics
stared at the clear night sky through a large eastward window
of the Albert Einstein Institute
at the patterns of the stars through the large window
She always felt awe as she contemplated
in that vast seeming hemispherical dome

She was startled by a momentary and unexpected streak of green light
just as the dawn was about to come upon her
mingled with some puzzling mathematical thoughts

that had been troubling her throughout the night
Then an odd thought overtook her...

This beautifully set story points out again the canon-wide theme for POPs that the awe of night skygazing is an experience fertile with new scientific understanding. For now, also note that the sudden and shocking perception of the pre-dawn green flash shifted Antea's experience of the sky.

And finally, let us listen to Lisa Randall's words. She quotes the Austrian mystical poet Rainer Maria Rilke to illustrate her relationship to the sublime (awesome) nature of the manifest universe, "For beauty is nothing but the beginning of terror, which we are still just able to endure, and we are so awed because it serenely disdains to annihilate us." She then adds her own poesy:

It is for these reasons both terrifying and compelling
science promise[s] to help transcend the narrow confines of individual experience
and allow us to enter—and comprehend—the realm of the sublime

Randall echoes the common POP declaration that awe at the manifest universe enhances scientific comprehension, yet she adds a new facet to the experience: fear. Entering into awe is accompanied by a sense of threat that the experiencer will "transcend" their individuality, a threat is felt that awe will "annihilate us." Just to foreshadow our model of awe, fear can accompany the experience of awe because, we suggest, awe arises as a liminal experience, occurring at the threshold of the super-psyche, on the "way in" to the realm of Nothing. The individual experiencer begins to lose identity (recall our discussion of the rarefying principle) at this threshold, and this rarefying is experienced as fear.

So, popularizer scientists universally acknowledge the experience of awe, and they proudly relate their own personal stories of awe and its effect on their career trajectories and new forms of scientific thinking. Yet, while allotting many words to the cause of explaining what the experience of awe is **not**, the POPs canon offers no elaborated reflection on the actual nature of the experience of awe. We will now do so. This reflection will culminate in Nihilology's model of awe that will serve as a powerful aid for the theorizing of popularizer scientists.

Now, the experience of awe, as we noted before, became secularized in the late eighteenth century; the experience was stripped of any divine involvement. Yet, this stripping did not spell the end of human fascination with the mysteries of awe itself. Indeed, this fascination has been reborn lately in the form of a social science redux as epitomized, perhaps, by the research of Dacher Keltner and his colleagues at University of California at Berkeley who study awe within the context of the psychology, sociology, and neuroscience of well-being. Scientists have actually become quite smitten by awe. What have they found?

Across the ever-growing research literature, there is agreement that three essential features characterize awe: perceptual vastness, extra-conceptual neural activity, and waning sense of self.

Researchers such as David Yaden, Michelle Shiota and Jordan Rosenfield, representing the fields of cognitive psychology, neural psychology, and history of science univocally conclude that awe is a complex emotional experience elicited by the stimulus of perceiving a vast or infinite space. The essential point for us here is that awe blooms from a visual event. It is a sort of seeing of space that introduces the scientist to awe.

We are reminded of the famous words of Albert Einstein when asked back in 1931 to give brief expression to his view of life, "It is enough for me to reflect upon the marvelous structure of the universe which we can dimly perceive... who can no longer pause and stand rapt in awe is as good as dead: his eyes are closed."

Secondly, there is no vital role for conceptual activity in experiencing awe. Indeed, research indicates that awe is "felt" by the autonomic nervous system when it is excited in a unique and uncommon manner. With most daily experience, one facet of the autonomic nervous system is excited while the other facet is quiet, yet during the awe-experience neural imaging shows that they are mutually excited at the same time.

Now, the autonomic nervous system is an involuntary, non-conscious network that receives information about the external environment and responds by regulating the heart, muscles, intestines, and glands accordingly. The sympathetic facet of the system speeds the heart, tightens blood vessels, and pumps up blood pressure ("Alert"). The parasympathetic facet slows the heart rate, increases gland activity, and eases sphincter muscles ("Relax").

We have many times called on the research of Andrew Newberg to throw light on the neural component of extraordinary human experience. Here is his perspective," .... intense stimulation of either the sympathetic or parasympathetic system, if continued, could ultimately result in simultaneous discharge of both systems (what might be considered a 'breakthrough' of the other system)."

So, awe is a non-conceptual "breakthrough" experience wherein the scientist is relating to their visual field under the sway of an autonomic nervous system that signals them to relax and be on the alert at the same time!

Now, we will plunge once more into neuroscience as a step toward elucidating our model of awe. For indeed brain scanning experiments have revealed that the parietal lobe of the brain is inhibited during the awe experience. The parietal lobe contributes to our spatial sense of self and orients our self physically, yet it appears to shut down during experiences of awe. The decrease of activity in the parietal lobe is thus accompanies by a loss of the sense of self.

The conclusion here is that the experience of awe is associated with a blocking of sensory input to the parietal lobe of the brain leading to a waning selfhood sense on the part of the awestruck. And further, some fascinating research has discovered a correlation between waning self and the sense of body size.

Psychologist Michiel van Elk and colleagues reported their findings in an article titled "Standing in Awe: The Effects of Awe on Body Perception and the Relation with Absorption." Their experiments were based on research findings that one's sense of the self is directly related to one's sense of physical body. Their study investigated whether awe directly affects body perception. And, indeed, it does. They

experimentally induced feelings of awe, and this resulted in higher ratings on the small-self-scale, consisting of items like "I feel small" and "I feel insignificant." van Elk then concludes, "Our findings suggest that the tendency to get absorbed in one's experiences predisposes people to experience awe and that a key feature of feelings of awe may be that people literally feel themselves as smaller compared to the environment."

Now, what do we make of this science of awe that concludes that awe arises upon the visual presentation of vastness, that it is an extra-conceptual experience indicated by a mutually excited sympathetic and parasympathetic nervous system, that it diminishes activity in the parietal lobe which correlates to a waning sense of selfhood accompanied by a sense of physical smallness in the awestruck?

Let us take a night-time stroll with a popularizer of science as they ponder a conundrum of physics. This scene is often reported in the POPs literature, and we have discussed it earlier. Lisa Randall, who focuses much of her attention on the creative act in science, reminds us of the state of the deeply focused scientist, one in a state where they can't-help-but-think-about-it. "...a scientist is likely to continue thinking about research when driving a car or falling asleep at night... Such embedded skills [of gaining insights] often continue operating in the background—even before they push good ideas into your conscious mind."

Before going deeper, let us note that Randall spends the entire concluding chapter of her book discussing creativity in science, yet presents no in-depth psychological model for how creative outsight occurs. She allows for unconscious parallel processing, as we discussed earlier, yet sidesteps any explanation for how it works and completely ignores any contribution from the super-psyche. As with most of her POPs colleagues, she is content to relegate these to the realm of the "un."

Now, when our scientist is on a night-time stroll while working intensely on an intellectual problem, other senses, including vision, do not function at a high level of perspicuity in that neural energy, and thereby awareness, is sapped by the mental quest. When the scientist takes the problem to the night sky, at first they are not actually looking at the dome of stars because there is no energy remaining for visual perception. We ask our readers to recall your own experience. When you are zealously pondering and driven to solve a problem, if you have your eyes open, you may not even notice what is in front of you. You are indeed thinking with your eyes open, but your attention is not on the information of light entering the eyes. You are looking, but you do not see.

This phenomenon has been well-understood for many years through research into the area known as "cognitive load" that tells us cognitive/sensory energy and awareness is limited in nature and intense attention to one mental or sensory task limits acuteness to other cognitive or sensory challenges. Beyond this basic interdynamic, there is one other research conclusion that has intriguing implications for our discussion of night-sky gazing. For, indeed, cognitive load is directly associated with dilation of the pupil in the eye. Dilation allows for, in fact demands, greater input of light through the eye to the brain and thus makes for better accuracy in seeing. Thus, when our scientist is intently involved in problem solving, their eyes are actually on deck for better night sky gazing.

So, for our scientist strolling on a clear night, two visual shifts must occur for them to enter into the experience of awe. One, the mentally absorbed scientist must shift their attention to the sky, the visual field that has been there all along. The scientist must become aware that the sky is there at all. This first shift is encouraged by the mental fatigue that sets in with long zealous conceptual concentration. Also, a shift away from awareness of thinking occurs, and a shift toward vision is stimulated by the fact that the cognitive load of attempted problem solving dilates the pupils of the eyes, allowing more light to enter and thus attracting our scientist to an awareness of seeing.

At this point, the sky becomes a trans-heuristic image, a tool for solving the problem at hand extra-conceptually. We have seen earlier how images have served throughout the history of science to solve problems that would not reveal themselves to thought or equations. Recall the imaging techniques of Crick and Watson, Roger Penrose, Richard Feynman, Steven Hawking, and Frank Wilczek.

When our scientist first becomes aware of the night sky, while conceptual work on their conundrum is beginning to ebb, yet with a certain amount of thought still occurring, there is no dimensionality to vision. The visual field is flat in our scientist's view.

This flat view of the night sky was perfectly described in the Christian Bible, Genesis 1: 6-19. God set a dome over the earth to separate the waters above from the waters below. The dome he called "sky." He then set stars in the dome of the sky to give light to the night. In other words, the sky is a two-dimensional curved disk, a half ball over the earth with tiny lights in it. The dome is called "firmament" which implies, from its etymology of "pounding out flat," that the dome is solid. This dome hung over the earth at a low enough altitude that birds could fly along its inner arch. Such a view held sway for only a short while as it was overturned by the earliest of celestial observations of the Greeks, yet it is representative of how we, unawares, perceive the night sky. We see a dome with lights pinned to it!

What is essential for our strolling scientist to become awestruck is a second shift in perspective, from an everyday (or every night) view of the sky as a pounded flat pot over the earth with lights affixed toward a vision of the sky's infinite depth, the depth of the universe as billions of light years of space. This expanse arrives concomitant with nervous energy being freed from the cognitive load of mental work and injected into vision. And when the three-dimensionality of infinite space enters experience, our scientist sees vastness. It is awesome.

The trans-heuristic potential of the night sky is often enhanced by a sensory surprise, as witnessed in the narrative accounts of POPs themselves. Recall the story before, related by Roger Penrose, of the sight of a "momentary and unexpected streak of green light, just as the dawn was about to come." Or the surprise sensation may be a bodily "tingling in the spine" as noted by Carl Sagan below. On one occasion for Rodney, the sensory surprise was aural in nature. He was standing on a second-story wooden deck gazing at the stars when, of a sudden, an acorn fell near his feet from a high oak tree. The hollowness of the wooden structure served as a drum. Boom! Awe at seeing the night sky.

So, the trans-heuristic value of the image of a dark, star-filled sky draws our strolling scientist away from cogitating and opens them to the vastness of seeing. It also aids them in going deeper into their conundrum by going outside of conceptual striving. This gives witness to the second major characteristic of awe: it is a deeply impactful emotional experience that does not include concept. (It is "felt" as witnessed neurologically by the mutually excited autonomic nervous system.) The third characteristic of awe also comes into play at this point: our scientist becomes aware of a waning selfhood. They feel themselves as small within the vastness of space. The self of the awestruck sinks away.

Listen to two more popularizers of science, their words in stanzas to emphasize the poesy.

Our feeblest contemplations of the Cosmos stir us
There is a tingling in the spine, a catch in the voice
a faint sensation, as if a distant memory, of falling from a height
We know we are approaching the greatest of mysteries
I believe our future depends on how well we know this Cosmos
in which we float like a mote of dust in the morning sky

This is Carl Sagan's voice calling out our human self as "a mote of dust." This statement is perhaps the most well-known from his bestselling book *Cosmos.* Echoing Sagan come the words of Michio Kaku from his article titled "It's Not All About You."

> Awe gives you an existential shock
> You are also dependent on something bigger than yourself
> Einstein was a tremendous influence on us
> because he was a messenger from the stars
> We look at the stars and think
> "My problems are so trivial compared to the majesty of the night sky"

So, we have thus far elucidated the three essential features that research findings of social and neural psychology tell us characterize our star-gazing scientist's experience of awe: perceptual vastness, extra-conceptual intense emotion, and waning sense of self.

Now let us leave our awestruck scientist alone with their deep emotion for the moment and remind readers of Nihilology's model of outsight that we presented earlier. For our perspective on awe is consistent with our model of outsight, and they combine to form a unified model of both awe and outsight. The unified model serves to bring us greater understanding of both experiences as liminal in relationship to the realm of Nothing.

Recall that at present, among popularizer scientists who contend with such issues, the most widely accepted model of insight was articulated by Bernard Lonergan back in the 1950s. His model consists of four stages:

1. Intensive thought on a problem
2. Conceptual rest
3. Insight
4. Verification

However, as a fruit of our review of the writings of the world's religions, the history of discovery in science, and research of neurotheology, we concluded that this four-staged model is superficial in that it fails to recognize the role of trans-heuristic images and it is resoundingly silent regarding the workings of the meta-conceptual. Lonergan relegates it to the realm of conceptual rest, the "un." He never actually steps over the threshold into the house of the meta-conceptual or the super-psyche wherein lives Nothing.

Thus, we suggested that scientific insight is more subtly and potently described as a five-staged process.

1. Intensive thought/employment of images
2. Conceptual rest
3. Meta-concept/Nihilitive knowing
4. A-ha! Outsight
5. Verification

In creating this five-staged model, we relied upon three unique notions contributed by Nihilology to the literature on scientific discovery: rarefying principle, being awake in the meta-conceptual/super-psyche, and outsight.

The rarefying principle allowed us to see a parallel process shared by science and religious practice. Science has applied the principle in its reductionist search for the essential makeup of the manifest world, concluding that this nature is more and more revealed by the shedding of excess matter. For religions, across history and the world, the rarefying principle traces the wearing away of selfhood on the path toward inperience of Nothing. The divine is encountered when the nature of the divine and the nature of the person are rarefied to Nothing.

In a wide-ranging discussion that took us through the history of scientific, artistic, and religious discovery, we learned that certain people, taking on specific modes of training, have become able to stay awake in the meta-conceptual. Others, with long practice in the disciplines of the world's religions, can even stay awake in the super-psyche where Nothing resides. Popularizers of science have shown little interest in either discipline, mainly because rational thought has no viability in those disciplines.

Finally, Nihilology showed why that which POPs commonly refer to as insights are actually outsights. They occur upon the **exit** from the meta-conceptual or super-psyche. Remember Lonergan, Wallas, and others all point out that a long period of intensive cogitation on a scientific problem leaves the scientist exhausted, and they begin a period of cognitive rest, perhaps musing on other topics, entering a reverie, and even snoozing. (Lonergan is satisfied to leave this period to the realm of the "un" whereas Nihilology holds this to be the interval of possible awakeness in the meta-conceptual and the super-psyche.) This period of rest ends when the scientist's thinking mind suddenly reawakens in a burst of energetic thought that solves their long-sought problem. This energy of outsight gives birth to elation. A-ha!

Now, by synthesizing the POPs narratives and the findings of awe science, that is, characterized by visual vastness, extra-conceptual emotion, and waning self-sense, **with** these three contributions of Nihilology (rarefying principle, awakeness in the meta-conceptual/super-psyche, and outsight), **within the schema** of the five-staged model of scientific discovery, we are able to create an all-encompassing six-staged model of awe and outsight.

1. Intensive thought/employment of images
2. Conceptual rest
3. Awe

4. Meta-concept/Nihilitive knowing
5. A-ha! Outsight
6. Verification

Recall our scientist out on a nighttime stroll. They are perplexed by a problem in their cosmological research and take it to the sky. The heuristic image of the dome of heaven aids them in suddenly seeing the true expanse of the vast cosmos. This vision quietens their thinking and the emotion of awe begins to arise in them. Being awestruck, their sense of self begins to rarefy as they step upon the threshold of the realm of Nothing. In that our scientist is not disciplined, their entrance into the realm of Nothing will be absent awakeness, carrying with it no inperience at all. With the return of conception, our scientist once again trods the threshold of Nothing and experiences the eureka! They then head for the lab to attempt to verify the new cosmological discovery.

Our new six-staged model of awe and outsight proposes that both are liminal experiences regarding the realm of Nothing. They arise on the threshold of the super-psyche. Awe arises on the ingress. Awe is what it feels like to approach the realm of Nothing. Outsight is what it thinks like to re-emerge from the realm of Nothing.

This model of awe and outsight is Nihilology's new proposal; it could become a helpful research guide into the science of awe. First, the model should be presented to the popularizers of science with the request that they report on their encounters with awe and outsight in light of the model's suggestions regarding sequence, liminality, and being awake. Furthermore, social science researchers, like Dacher Keltner and his colleagues at the Greater Good Science Center, should design experiments that explore the relationship of awe and outsight. As of yet, they have not been approached as a liminal pair.

Let us close this phase of our theme by making an observation in the same vein as those offered in our earlier discussions regarding outsight. According to Nihilology, popularizers of science take the rewards of (unawake) meeting with Nothing while denying its reality. POPs allow for awe in bringing them into the presence of the unthinkable universe. Therefore, we suggest that they must, by force of rational argument, also allow for the actuality of the meta-conceptual and super-psyche. In our six-staged awe and outsight model, we contend that POPs do indeed delve into the realm of meta-concept and super-psyche, but without being awake therein. They lack in practice and lack in discipline. In the end, POPs are not motivated to become more sophisticated in their understanding of awe because it would undermine their *heimwelt*, their intersubjective truth of a naturalistic universe, a limited realm of concepts and senses.

Our observation that POPs may not be awake to stage four of our model of awe and outsight is well-illustrated by a profile presented earlier of theoretical physicist and Noble laureate Frank Wilczek. He is enamored with the Great Symbol of Taoism, the yin yang sign, both as thing of beauty and as a heuristic image that reveals much about the relationship of substance and force in the manifest world. This relationship is one of complementarity, which demands that we see one thing from two perspectives in order to expose the truth of the thing. Now, Wilczek's shortcoming in his vision of the yin yang sign is that, while being sensitive to the many connotations of the two swirling fish, he is not awake to the fact that the outer skin of the two fish forms the circle in which they swirl; he misses Nothing.

So, in this theme we bring to light certain specific aspects of the professional life of the popularizers of science of which they themselves are unaware, or at least they are unwilling to acknowledge. POPs exercise a form of faith, and they make use of the heuristic power of faith to ply their theoretical and empirical trade. POPs experience—and are openly proud of their experience here—awe and outsight. Both are liminal regarding Nothing and only come to be as dependent on Nothing.

Now, let us add one final aspect to this list, for indeed POPs have begun to re-introduce the human person back into their canon and thereby have chiseled a chink in their objectivity, in the bedrock of the scientific worldview.

We discussed earlier how natural philosophers were motivated to understand the manifest world in order to reveal, and revel in, the majesty of divine creation. Over time this motivator lost its spirit; research went forth toward understanding alone without any laud for the divine. Science as we know it today was born. Also, being inheritors of the Enlightenment, scientists held that the tool for this free-standing knowledge of the world is rationality alone. Thus, the scientific world was stripped of a role for the divine creator and stripped of a role for divine revelation. **And** science stripped one more thing from the world: the central role of the human being.

The position that human beings are a happenstance in the cosmos, one among an infinity of creations, is, even today, held to rather stridently by POPs who align themselves with a group of scholar philosophers known as the New Atheists. Lawrence Krauss, physicist at Arizona State University and best-selling popular science author, is leader among them.

In a lecture to employees at Google, Inc., highlighting his most recent bestseller *The Greatest Story Ever Told... So Far: Why Are We Here?* Krauss imagines religion and physics in their at-odds approaches to explaining human life on an ice crystal. He opines that religions would argue about which way to face on the ice crystal in order to pray and fight wars over which was the true way. Of course, this fight is based on seeming antagonism. The ice crystal could face in any direction. Krauss then brings down the gavel, "And that's essentially the world we live in. All of the facets of the world that make it look like it was designed for our existence are pure accidents in a fundamental way."

Coincident with this position, as held by Krauss, comes the belief that the universe is perfect. Recall our earlier discussion of the basic flaw in the manifest world where we made it clear that the world's religions teach the lack regarding the human being: illusion, ignorance, sin. The reason why Krauss sees the universe as perfect is because humanity is cosmologically irrelevant, a chance occurrence which is unessential to the basic physical reality of our universe.

Now, while this stringent position perhaps remains the ideal for a few popularizers of science, it is involuntarily beginning to bend and curve by force of the scientific theories espoused by the POPs themselves, namely the Heisenberg Uncertainty Principle and the Anthropic Principle.

Before entering into our analysis of these principles and their profound effect on the status of humans in the universe, let us recall that, in the centuries when the worldview of natural philosophers was being usurped by science, three things were stripped from the old worldview: a role for the divine creator, a role for divine revelation, and a central role for the human being. Directly implied by these three strippings was a fourth; the emotional, believing, subjective, and narrative components of the human being became regarded as of no consequence in the **task** of science. Doing science is objective.

The *Stanford Encyclopedia of Philosophy* gives a typical definition of scientific objectivity, "an ideal that the claims, methods and results of science are not, or should not be influenced by particular perspectives, value commitments, community bias or personal interests... Objectivity is... the basis of the authority of science in society."

However, recall how we pointed out at the top of this section that popularizers of science, in their usurpation of the role of religion in society, have taken on the function of supplying the mythic stories of our modern scientific worldview. Having opted to rid their canon of mathematics, they have become more and more reliant on literary techniques to convince their lay public. In doing so POPs have become storytellers, and in taking on this role, they have re-introduced their personhood into the task of science. We might say that old school academic science writing was typified by third person passive constructions, "It was concluded... " POPs science, in contrast, is typified by first person active constructions, "I realized..."

"My goal... is to invite you along on my personal journey of exploration... this book is scientific autobiography of sorts... it's my personal quest for the ultimate nature of reality... "

As an example of this first-person writing, we quote once again Max Tegmark, a POPstar featured in dozens of science documentaries, from his bestselling book *Our Mathematical Universe: My Quest for the Ultimate Nature of Reality*, "My goal... [is]to share with you my excitement and reflections about these mind-expanding mysteries... . it's certainly not your standard popular science book that attempts to survey physics in an objective way..."

Tegmark allows, without professional embarrassment, that this book of popular physics is not written "in an objective way," an admission that would have gotten him kicked out of the lecture hall not too many years ago. Thus, according to the POPs culture, it is good science to tell personal stories and exhibit one's excitement as a scholar. And Tegmark is only the pinnacle of an established tradition in this first person writing. Recall Carl Sagan's *Varieties of Scientific Experience: A Personal View of the Search for God* and Francis Crick's *What Mad Pursuit: A Personal View of Scientific Discovery.* This tradition speaks clearly to us as readers; the POPs worldview is holding the door ajar for the scientist herself/himself to enter into its realm of study. Persons are finding their way back into legitimate science.

So, now that we see clearly the POPs have insinuated their personhood into the task of science, their narrative subjecthood into their story telling, we can take a look at the actual science they are telling stories about. First, the Heisenberg Uncertainty Principle, which is ever-present in the canon of popularizers of science.

When Danish physicist Werner Heisenberg was only 26 years old and seeking position in the academic circles of Europe, he had an outsight that would earn him a Nobel Prize and shift the vision of science regarding the nature of the physical world. Simply put, he noted that a scientist seeks to measure certain properties of an electron by illuminating the particle with light or radiation, yet the energy of the light itself influences the electron. For example, to get a bead on the electron's position, the scientist must beam short-wave, high-energy light which alters the momentum of the particle. To get a bead on the momentum of the electron, the scientist must beam long-wave, low-energy which it makes its position more hazy. Thus, and this is the outsight, a scientist cannot measure the position and momentum of an electron simultaneously.

First of all, let us note that Heisenberg came up with his outsight by employing a heuristic image, in a manner akin to so many scientists in history. The gamma-ray microscope that his thought experiment employed did not exist in his day. Rather, following a long period of intensive thought with Niels Bohr, he had a period alone to muse. He envisioned a microscope shooting out rays that bump off electrons and reflect back to the eye of the observer. A-ha!

Early on, Heisenberg was unsure and modest about his discovery, speaking in terms of "uncertainty relations" and "indeterminacy" in measuring electrons and other, what science now call, quantum particles. Yet, over time and enhanced by the fervent input of fellow scientists, his outsight became widely known as the Uncertainty Principle, and the implication of his principle evolved into an examination of how scientists may influence the natural world by observing it.

The seed of this existential predicament was already planted in Heisenberg's 1927 paper. "I believe that the existence of the classical 'path' can be pregnantly formulated as follows: The 'path' comes into existence only when we observe it." The classical path is the world according to Newtonian science, what we laypeople call our daily world. Heisenberg's seed is actually a flourish; our daily world only exists by us looking at it! How can this be?

In 1958, Heisenberg published his classic work *Physics and Philosophy: The Revolution in Modern Science.* In our minds, the clearest précis of his argument therein is offered by Paul Davies in the introduction to the 1962 edition. Davies is a leading voice in the POPs movement and, as the director of BEYOND: Center for Fundamental Concepts in Science, is an expert voice in the large ideas of physics.

We should note, as a foreshadowing of our next discussion, that Davies is a deep thinker regarding the ramifications of the Anthropic Principle as it applies to the re-introduction of the human being into scientific worldview. For example, in the preface to his best-selling book *The*

Mind of God: Basis for a Rational Universe, he expounds the belief, "... I have come to the point of view that mind... is not a meaningless and accidental quirk of nature, but an absolutely fundamental facet of reality... we human beings are built into the scheme of things in a very basic way."

Now, regarding Heisenberg's main theme of the Uncertainty Principle, Davies asks the primal question as to the nature of an electron. In response, he holds that it is at base not a physical thing at all, rather a set of potential outcomes of measurement. Its reality lies in its observation, not in the electron itself.

Heisenberg himself puts forth this basic uncertainty in a bit more poetic style, quoted here in stanza:

> In the experiments about atomic events we have to do with things and facts
> But the atoms or the elementary particles themselves are not as real
> They form a world of potentialities or possibilities rather than one of things or facts

From this we learn that quantum particles, electrons for instance, do not exist as material particles with defined properties until **after** they are pinged. Before measurement the electrons exist in an undetected, indeed undetectable, state, waiting for a signal from the material plane to manifest a certain one of its potentials. The ping or the energy of measurement, creates what scientists know as an electron.

The idiosyncrasies of polarized light may help clarify how it is possible for scientists to know that something can exist in an undetectable state.

It is well established that visible light is a type of electromagnetic wave that flows in one of four directions relative to the observer: horizontal, vertical, diagonal, or circular. A typical beam of light is comprised of a random assortment of electromagnetic waves in the first three directions. Circularity is created artificially by shifting one of two orthogonal beams of light by 90 degrees.

Now, imagine in your mind's eye the rays of the sun being focused through a particular type of optical filter called a horizontal polarizer. It will only allow the horizontal waves through and cancel the vertical and diagonal ones. Then, when the rays that have exited the horizontal polarizer are focused through a vertical polarizer, all rays are canceled. No detectible light exits the second filter. However, a interesting phenomenon reveals itself if a diagonal polarizer is set up between the horizontal and vertical filters. Our image demands if a filter is **added** there would still be no light exiting the threesome of filters. But that is not what happens. In fact, light does exit when there are more filters! Where does it come from?

The phenomenon is explained by what scientists refer to as "superposition" which is the capacity of a quantum system—in this case photons of light—to exist in multiple states until it is measured. The horizontal light exiting the first filter is in superposition with respect to the diagonal filter which measures it, thereby collapsing it into diagonal light. Thus, there exists some amount of diagonal polarized light focused on the final vertical filter. Then half of the diagonal wave will make it through the vertical filter. A-ha! Visible light.

Astonishingly, not being able to measure light does not imply there is no light. We are not talking about some faint signals that are detectable in theory, but that technology is not evolved enough to measure. Rather, if no signal can be found on the material plane, if the light is not anyTHING, there can still be electromagnetic energy on some undetectable plane waiting for the appropriate "polarizer" to stand between the experimental setup to reveal it.

So, this simple experiment helps us understand that, on the quantum level, the very existence of a thing is brought to be through observing it. And, via extension, it reveals a basic notion of Nihilology as well. Because an unmeasured electron is undetectable does not mean it is Nothing. Rather this "not anyTHING" is a wonderfully fecund nothingness.

One more point regarding scientific thinking about quantum particles is inherent to Heisenberg's statement, and Davies's interpretation, quoted before. Quantum theory is epistemic in nature, a theory about what human beings can know. It proposes to explain how we observe quantum phenomena the way we do. It tells us about the way the human mind works. It is a theory of knowledge. And knowledge inherently implies a knower. The human being, by observing the world, plays an essential role in making the world real!

Professor Vuk Uskoković from the University of California at San Francisco reflects on such ethereal matters in an article titled "The Metaphorical Model: The Bridge Between Science and Religion." He states that there is no viable means to make conclusions on any aspect of physical reality without taking into account the observer of that reality. He employs the phrase "co-creation of human experiences."

Heisenberg wrote in a similar vein, "...in natural sciences, the object of research is not any more nature in itself, but nature exposed to our method of questioning, and in this sense, man herein faces himself."

Thus, Heisenberg's Uncertainty Principle, in pointing out the potentiality nature of quantum particles before measurement and the essential role of human observation in the creation of actual material particles, does not allow for classic objectivity in the pursuit of scientific knowledge. Human beings, in all their emotional, believing, subjective, and narrative components are essential to the task of science and to the object of scientific study. And the popularizers of science are eager leaders in exploiting this fact as they bring their canon to the public.

Now there is one other principle in the POPs canon that is akin to the Uncertainty Principle and even more dramatic in its effect on the objectivity of the popularizers of science. It is known as the Anthropic Principle. In its etymology alone, we should be alert that this principle re-introduces the POPs personhood back into the world in some fashion. The word "anthropic" derives from the Greek *anthropikos* which means "for the human."

In 1973, Australian physicist Brandon Carter gave what history now knows as a eureka! speech. On the 500th anniversary of Copernicus's birth, he introduced the world of science to a new vision of the human in the universe. (It is, of course, an aged vision in the world of philosophy and religion.) The Copernican Principle had held sway for centuries; it states that human beings hold no privileged position in the cosmos. Carter reflected in his well-known modest statement, "Although our situation is not necessarily *central*, it is inevitably privileged to some extent." He tagged this as The Anthropic Principle.

Carter was speaking to a cosmologist's conundrum. To wit many of the accepted physical constants in the universe, like the proton's mass or Newton's gravitational constant, only hold true at this time in the history of the cosmos. These constants are necessary for the birth of intelligence, so it appears we humans are living at a special instant in time, because at any other instant there would be no consciousness to witness the constants.

Roger Penrose, renowned British mathematical physicist, wrote one of the most influential books in the POPs canon, *The Emperor's New Mind: Concerning Computers, Minds, and the Laws of Physics.* It won the Royal Society's Science Book Prize in 1990 and has been reprinted many times in the decades since. Penrose asked these pithy questions regarding The Anthropic Principle, "How important is consciousness for the universe as a whole? Could a universe exist without conscious inhabitants at all? Are the laws of physics specially designed to allow for the existence of conscious life?"

As we read through his elaborations on these questions, it is clear which answers apply in his estimation. Human consciousness is important to the makeup of the universe. No, the universe could not exist without conscious inhabitants. Yes, the laws of physics are such that human consciousness comes to be.

Many POPs write about The Anthropic Principle, each from their own personal perspective. (Though usually without coming to the obvious conclusion that scientific objectivity is sullied in the writing.) Max Tegmark, for example, believes that humans lend meaning to the cosmos. In a blog post titled "We're not insignificant after all" he makes a personal guess that humans are the only life in the universe advanced enough to build complex machines like telescopes. He then puffs up, "... those galaxies are visible and beautiful to us—and only us. It is only we who give them any meaning..."

Now, perhaps the most aesthetic cosmic essence among the POPs canon is suggested by Frank Wilczek in his book *A Beautiful Question: Finding Nature's Deep Design.* He perceives a world where the mind of the human being is by nature one of "miraculous harmony" with the material world. This harmony finds its roots in beauty. "When we form expectations that turn out to be correct, we experience pleasure and satisfaction. Those reward mechanisms... stimulate—indeed, at base they *are*—our sense of beauty."

Frank Wilczek argues that beauty is embedded in the natural world and in the human being. Self and object share in a quality of beauty, so when we discover that *our* sense of beauty is embodied in the physical world, we are discovering something about us as well. He then gives the example of co-embodied light. "Our sense of vision... [is] conditioned by our interaction with light... That ability is hardwired into our brain... In the more advanced, modern parts of physics we learn that light itself is a form of matter, and indeed that matter in general, when understood deeply, is remarkably light-like."

We presented earlier that POPs have a belief in beauty, that beauty is a guide to understanding the universe. A belief that echoes the old detective movie saying, "Follow the money." Well, POPs are following the beauty. Now, in the voice of Wilczek, we hear a new resonance to this faith in beauty. The universe is lawed in such a beautiful way that these laws are inevitable. The world could not be in any other form. So beautiful.

Further, Wilczek holds that human beings are innately linked with the world. Both the human mind and matter are at their core beautiful. The human mind participates in the world as beauty being appreciated by beauty. We would like to point out that this is a step toward the ultimate realization of Nihilitive Knowing where self as Nothing greets the nature of the world as Nothing. Wilczek takes this step in his discovery of the likeness between the human observer and the world observed.

We noted before that, in the worldview of religion and philosophy, the import of *anthropos* has long been recognized. Ray L. Hart's book *God Being Nothing* makes clear statements regarding the three Abrahamic faiths' essential primacy of the Anthropic Principle, "Each of the three religions valorizes God, the world, the human creature, and their interrelation. Remove the human creature and there isn't much left from the perspective of human meaning."

From our discussion of The Anthropic Principle in the POPs canon, it seems that popularizers of science have come to validate a basic belonginess of humans in the cosmos akin to that which has been taught in the religious worldview for a long time.

So let us reiterate the crux of our argument in this theme. First, The Heisenberg Uncertainty Principle asserts that human beings, in their observation of quantum particles, affect the manner in which particles express as physical entities, and therefore human beings are essential to the creation of the physical universe. Thus, Heisenberg left the door ajar for the popularizer of science to be re-introduced as an essential aspect of the scientific endeavor, causing a chink in the bedrock of scientific objectivity. Second, The Anthropic Principle asserts that our universe is lawed in such a way that only humans can be aware of it, that this lawed world is made for the human being. This principle throws the door wide open. POPs are rushing into the scientific endeavor in the uneasy understanding that scientific objectivity may indeed be impossible, indeed may be unscientific.

Let us then conclude this section. It is our contention that popularizers of science don't need us to argue them into accepting religious approaches to understanding the world we live in. They are already doing religion!

POPs show a deep faith that the universe is of a unified nature, fully explainable by a grand unified theory. They also benefit from the heuristic power of faith in beauty to lead them toward this holy grail of a single theory of everything. Further, POPs also proudly extol their experiences of awe as giving birth to their careers in science as well as to new paths of research inquiry. And, as we noted, until its secularization in the late eighteenth century, awe was universally accepted as an experience that revealed the divine.

Going deeper, in elucidating Nihilology's unique model of awe, we established that awe is a liminal experience regarding the realm of Nothing, the realm of the divine, and that awe is fully dependent on this relationship with the realm of Nothing. Awe would not exist were it not for Nothing, not for the divine. Finally, by reintroducing persons into science, both in the act of composing the myths of the scientific worldview and in the world studied by science, POPs, by definition, have chiseled a chink in their sacred objectivity, allowing, as religion has ever done, the emotional, believing, subjective, and narrative components of the human being to be an integral aspect of the cosmos.

Scientific Theory Evolving into Philosophy

We concluded the previous theme by stating that popularizers of science are already doing religion. They don't need to be argued into it. Just so, we will conclude this theme by stating POPs are already doing philosophy. Their own canon proves this conclusion to be valid.

First, let us put forth three heuristic statements—guiding rules of thumb—that will help us navigate this discussion. We will also substantiate the power of these rules of thumb as we move through our argument.

1. Theology is the search for understanding a universe with God in it. Philosophy is the same search in a universe with no God in it.

2. Science seeks to understand the universe via the "how?" question. Philosophy is the same search via the "why?" question.

3. Scientific discovery, at various epochs of history, has been led by empirical experimentation, by mathematics, or by theory. When led by theory, science is indistinguishable from philosophy.

Keeping these rules of thumb on hand, we can learn much from the statements of the POPs themselves as they frequently and openly opine on the liaison between science and philosophy. For example, Stephen Hawking often said so, and millions of people around the world were listening. Hawking was the Director of Research at the Center for Theoretical Cosmology at Cambridge University. He has been ranked among the top 100 Greatest Britons. We noted earlier his super-nova book *A Brief History of Time* that holds the record for weeks as a *Sunday Times* best-seller. When Hawking spoke, he was compelling our view of the world.

With this in mind, listen to these words from his 2010 Amazon topper of the sales ladder *The Grand Design*. First Hawking asks the classic questions of where the world came from and if there is a need for a creator in this coming to be. He then states, "Traditionally these are questions for philosophy, but philosophy is dead. Philosophy has not kept up with modern developments in science, particularly physics. Scientists have become the bearers of the torch of discovery in our quest for knowledge."

We could not help but hear, in this bold dismissal by Hawking, the echo of Friedrich Nietzsche's "Gott ist tot," first spoken by Nietzsche in his early work *The Gay Science* and made famous as intoned by his maniacal persona Zarathustra. And a reading of Nietzsche's famous dismissal sheds much light on what Hawking intended here.

Regardless of how anti-intuitive this sounds, Nietzsche was not making a metaphysical statement with his "God is dead." Recall from our earlier discussion, Nietzsche (and Sartre), averred that there are only phenomena, no real hidden beneath or above the daily world of common perception and reflective thought. This hidden, or noumenal, is the fantasy foisted on the peoples of Europe by philosophers and Christian priests. Thus, "God" is a not an ontic being who once was and now is not. Rather, Nietzsche was intoning the demise of the **image** of God and the **beliefs** surrounding that God. He was announcing that the faith held in his mid-nineteenth century Europe had become dysfunctional. It had lost its heuristic value and power to guide people's lives.

Nietzsche was a watcher and interpreter of the nihilistic phenomena that was seeping into European culture, specifically into the German people and the German language. Note this statement from his posthumous work *The Will to Power*, "Nihilism represents a pathological transitional state—what is pathological is the tremendous generalization, the inference that there is no meaning at all."

So, Nietzsche was an observer of the nihilistic culture around him. Now, what did he observe about God? "Whither is God? ...I will tell you. We have killed him, you and I. God is dead. God remains dead. And we have killed him... Must we ourselves not become gods simply to appear worthy of it?"

Hawking was making the same observation here with his "philosophy is dead." He took a penetrating look at the state of philosophical investigation in the West. He concluded that it has lost the heuristic power to reveal the underlying nature of the universe and could no longer guide people's lives therein. Thus, he said to the community of POPs, "We have killed it, you and I. Philosophy is dead."

Hawking believed that the POPs' way of revealing the nature of the universe has usurped the way of traditional philosophy that he and his cohorts have killed; scientists are the new philosophers.

(Hawking also assumed the persona of übermensch in today's western culture, just as Nietzsche did in his day. This persona has a detrimental effect on science discovery, as we discuss below.)

Now, it is important for readers to understand that Hawking was not making a metaphysical statement with his "philosophy is dead." Philosophy is not an ontic enterprise that once was and now is not. It is the beliefs and images of philosophy that he and his POPs cohorts killed. For indeed, Hawking eagerly takes on the "ontic" philosophical enterprise, in its new façade as popularizing science. Recalling the first two of our heuristic tests for the presence of philosophy, Hawking seeks understanding of a universe with no God in it and is guided in that search by the "why?" question.

To wit, we turn to the introduction to his book *The Grand Design* wherein he intones his belief that the deepest understanding of the universe results from asking how the universe goes about its business in conjunction with asking why it goes as well:

Why is there something rather than nothing?
Why do we exist?
Why this particular set of laws and not others?
This is the ultimate question of Life, the Universe, and Everything

These are words of philosophy. Also, note how Hawking capitalized "Life," "Universe," and "Everything." Revealing the exact same ambivalence that we noted before about Carl Sagan and Stephen Hawking, while stating boldly that there is no such entity as a personal divine creator, personalized the world in the "person" of Life. Further, Hawking formatted the philosophical questions into a stanza, expressing the notion that poetry is more suitable to philosophizing about the nature of the world than essay-formatted paragraphs.

So, let us then harken to the words of five other stars among the popularizers of science to conclusively establish our argument that POPs are modern-era philosophers. First, make note of the recent *New York Times* bestselling book by Sean Carroll, a theoretical physicist from California Institute of Technology: *Big Picture: On the Origins of Life, Meaning, and the Universe Itself.* Indeed, Carroll's philosophical approach greets readers without their even cracking the book's cover. The "why?" question inherently is a question of meaning, and Carroll inserts "meaning" into his very title. So, he unashamedly addresses "why?" and is, nonetheless, accepted into the POPs canon as a leading voice.

Life is not a substance, like water or a rock
It's a process, like fire or a wave crashing on the shore
It's a process that begins, lasts for a while, then ultimately ends

We have taken the liberty of setting these words from Carroll's preface in stanza format to visually emphasize the poetic nature of his philosophical language of why. In fact, we have not taken undue license with Carroll's writing, for Carroll himself terms his approach to theoretical physics "poetic naturalism."

Continuing in the preface, Carroll opines that in the old way of scientific thinking, human life couldn't possibly be meaningful because humans were just a collection of atoms swirling about under sway of the laws of physics. But, according to his POPs way, humans "are collections of atoms, operating independently of any immaterial spirits or influences, *and* we are thinking and feeling people who bring meaning into existence by the way we live."

So, Carroll believes that the universe is one of physical laws personalized by why-asking human beings. ("By the way we live" is an attitude of a sort of philosophy called Existentialism, the sort of philosophy that POPS adhere to, as we shall discuss soon.)

Brian Greene, in his book *The Elegant Universe,* while assuming our first and second rule of thumb regarding the presence of philosophy, is especially forthcoming in his avowal of our third heuristic statement. Science led by theory is indistinguishable from philosophy. He points out an essential debate going on in the world of science regarding how science **should** be done. "The 'traditionalists' want theoretical work to be closely tied to experimental observation, largely in the successful research mold of the last few centuries. But others think that we are ready to tackle questions that are beyond our present technological ability to test."

Now listen to the voice of American physicist Alan Lightman, best known throughout the world as the writer of the novel *Einstein's Dreams.* He is also the author of multiple books that have built the POPs canon. Here, in his preface to Alan Guth's book *The Inflationary Universe: Quest for a New Theory of Cosmic Origins*, he provides a nice overview of how science has become the cultural bearer of "why?" He notes that throughout history peoples have had their own unique cosmology, their own story of the birth of the universe, but that in the contemporary West cosmology has emerged as a science, a research specialty within astronomy and physics. "In the 1970s... cosmologists began seriously asking questions like: Why does matter exist at all... Why is the universe as homogeneous as it is over vast distances?... In other words, the nature of the questions changed... 'Why?' was added to 'what?' and 'how?' and 'where?'"

In our modern culture, cosmology has shifted from the purview of religion to that of science, culminating an evolution that began back in the sixteenth century. Yet, now we see the evolution continuing with the introduction of the "why?" question; science is widening its scope to incorporate the quest for meaning in the universal manifestation. POPs science now, by its very nature, is its own form of philosophizing.

Now, while exploring the POPs canon, we have discovered that in their prefaces and epilogues, in their beginning and ending words, we often find revealed the overarching perspective and beliefs of the authors. For example, this is the culminating statement of Guth's book. It shows his philosophical dreams for the profession of physics, pointing out that the attempts to describe the materialization of the universe from nothing represent an exciting expansion of the boundaries of science. "If someday this program could be completed, it would mean that the existence and history of the universe could be explained by the underlying laws of nature. We would have accomplished the spectacular goal of understanding why there is something rather than nothing..."

From all of the above we conclude that popularizers of science are indeed doing philosophy. They research a universe with no God in it. They ask of this universe "why?"; their mathematics and empirical experimentation are led by theory.

And finally, listen to the words of David Z. Albert, director of the MA Program in the Philosophical Foundations of Physics at Columbia University, in an interview titled "What is the Role of a Philosopher or Science." He recounts that his Ph.D. is in theoretical physics and that he taught in physics departments before he taught in philosophy departments. "My rule is now, when I write a paper, if in the end it has more than two equations, I send it to a physics journal. If it has less than two equations, I send it to a philosophy journal."

We discussed earlier that the absence of mathematics, the lingua franca of academic physics, is a decisive key to differentiate traditional academic science from popular science. POPs books rely on stories and images and metaphors to instruct their audiences. They often openly declare that they scrub their presentations of mathematics in order to speak to educated lay readers. While the absence of mathematics is not a rule of thumb for the presence of philosophy, as that would imply every novel or essay is philosophy, these words of Albert do make us wary. If a book of **physics** has no mathematical equations, there may be some philosophizing in the pages ahead!

Albert then continues with a perspective that serves well to cap this phase of our discussion of POPS as philosophers. He concludes, with Nihilology, that the domain of physics and the domain of philosophy are not necessarily bordered, that when the domains are without border guards it is a sign of health for both, "I think Philosophy of Science is at its best and most exciting at historical moments when it is not so easy to distinguish between certain kinds of theoretical physicists and certain kinds of philosophers... Whether these problems properly belong to physics or they properly belong to philosophy, when the field is healthy, isn't much of an issue."

We, as authors, have long wondered and reflected on the reason behind popularizers of science taking this path that inherently expands the borders of science to incorporate the domain of philosophy. For us, the reason has to be that POPs themselves are asking questions that are bigger, faster, and more primordial than their traditional tools are capable of measuring. They cannot resolve these questions and have thus reached to philosophy for aid in doing so. This path is, of course, the way of Nihilology which holds that in the **pattern** of philosophy, science, and religion is found the answers to bigger, faster, more primordial questions about the universe.

Before moving on from this phase of our discussion, it must be acknowledged that there are one or two outliers among the community of popularizers of science. Recall that Lawrence Krauss holds stringently to the position that humans are a mistake, a happenstance among the infinite creativities of the cosmos (a belief that is bending and curving among most of his POPs colleagues). Well, Krauss also holds stringently to the belief that science is not and should not be plying its trade in the realm of "why."

Here is a typical quote from Krauss's first best seller *A Universe from Nothing: Why there is Something Rather than Nothing*, "When we [scientists] ask "why" we usually mean "how?" If we can answer the latter, that generally suffices for our purposes... "Why" implicitly suggests purpose, and when we try to understand the solar system in scientific terms, we do not generally ascribe purpose to it."

Krauss then proceeds to explain the subtitle of his book as really intending to ask **how** is there something rather than nothing, for according to his tight view, "how" questions are really the only ones that can provide definitive answers via the scientific study of the natural world.

Krauss is unique among POPs in his sharp-edged belittlement of philosophy. He often speaks of ageless questions as being transformed and brought closer to being answered by science because "empirical knowledge shines new light on otherwise dark corners of our imagination."

Those dark imaginings are, of course, philosophy. Another favorite criticism, employed so many times he calls it his personal beating of a dead horse, is to state that science frames questions in the mode of "how," which produces true and useful new knowledge, unlike the "why" questions of philosophy "which generally presume the answers up front."

Yes, Krauss holds to his "I don't do philosophy" stance. However, Krauss is hypocritical in this stance as pointed out by many leading voices in history and philosophy of science. Listen here to the logic and clarity of George Ellis, a South African physicist and lauded authority on physical cosmology upon being asked if Krauss had solved the mystery of how there is something rather than nothing. "Certainly not. He is presenting untested speculative theories of how things came into existence... It's a philosophical speculation... It's very ironic when he says philosophy is bunk and then himself engages in this kind of attempt at philosophy."

To conclude then, POPs are engaging in philosophy in their pursuit of cosmological and other theoretical physical answers. That is, they search for understanding a universe with no God in it, they often search via the "why?" question, and their search is often led by theory. Having established this, let us go one final step to show that POPs are exhibiting clear affinities with the two specific schools of philosophy that we discussed earlier: Existentialism and Nihilism.

Recall from our in-depth discussion before that for Existentialists, as witnessed in the works of Jean-Paul Sartre, there is no real hidden under the daily world of common phenomena and reflective thought. The so-called noumenal realm with its divine personage is a fantasy of misguided philosophers and priests. As Sartre (and Nietzsche) joked, God's one essential downfall is that he doesn't exist! The human dilemma, for the Existentialist, is that we sense our very existence is superfluous (*de trop*); there is no floor beneath us and no divine guidance for our days on earth. This very sense defines Existentialist nothingness.

Now, popularizers of science often exhibit a not-so-subtle superior attitude toward their lay readers by chiding them for reaching for God to supply buoyancy as soon as they feel this "no-floor-beneath-their-feet." This misguided tendency, the POPs hold, results in a "God of the gaps." As Brian Greene says, "What science is pretty good at ruling out is the so-called "God of the gaps"—the traditional way of invoking God whenever there's something in science that we haven't figured out."

Neil deGrasse Tyson, American astrophysicist, author of *Astrophysics for People in a Hurry* (*New York Times* bestseller), and POPstar expands upon Green's words. In a recent interview, Tyson opined that over the long history of scientific discovery there has always been, and still is, a frontier beyond which lies the unknown. Further, it has been the bent of religious people to posit the existence and handiwork of God just beyond that boundary. "This, over time, has been described by philosophers as the God of the gaps. If that's where you are going to put your God in this world, then God is an ever-receding pocket of scientific ignorance."

Again, this gaps notion reveals a superioristic attitude on the part of POPs, as if people who have faith in God only find him in the dark recesses where science has not yet projected a light. As if scientists hold sway over just where God can be found! Nonetheless, this belief on the part of POPs that no divine person exists is consistent—even in the attitude—with Existential philosophers.

A universe without the presence of God is the norm throughout the POPs canon. (Recall our first heuristic rule of thumb for detecting the presence of philosophy.)_For example, Stephen Hawking was famous for stating that a grand unified theory would reveal to scientists the very mind of God. When pressed to explain this audacious statement, he replied, "What I meant by 'we would know the mind of God' is, we would know everything that God would know, if there were a God. Which there isn't."

We discussed earlier how POPs use religious language irresponsibly to draw in their lay readership. This "mind of God" statement by Hawking is yet another example of such language.

One more instance, Lawrence Krauss, in the very first line of his book *A Universe from Nothing,* shows how important it is for him to deny God, just as it was for Sartre and Nietzsche. He "admits" that he is unsympathetic to the widely held religious conviction that creation requires a creator. Then, near the end of that same volume, Krauss puts forth a classic Existentialist statement regarding belief, "[S]cience... makes it possible to not believe in God... Religious belief... becomes less and less necessary, and also less and less relevant... No doubt... God is unnecessary—or at best redundant."

As we foreshadowed earlier, readers certainly hear in these words of Krauss the echoes of Jean-Paul Sartre from his famous speech "Existentialism is a Humanism." Let us quote Sartre once again, "... if God existed that would make no difference from our point of view... the real problem is not that of His existence; what man needs is to find himself again and to understand that nothing can save him from himself..."

Neither Krauss nor Sartre feel compelled in their philosophizing to once and for all disprove the existence of God. For them to succeed in plying their trade, God must be, well, beside the point. God is unnecessary, redundant, not the real problem, and not worth the effort of disproving.

To go one step farther in this discussion, we should also heed others of Sartre's words, "man needs to find himself again and to understand that nothing can save him from himself." Sartre holds fast to the recognition that humans are on their own with no God as their creator or guide. Each individual must assert their essence. An authentic life is created by that assertion. Nothingness, for the Existentialist, is the bleak darkness that reigns underneath this assertion, making human life one of nausea and anguish. The only achievable human goal is to live as an independently acting, responsible, reflective person. This authentic life is the self-creator of meaning, "Man is what he wills... nothing else but that which he makes of himself." "Life has no meaning *a priori* ... It is up to you to give it a meaning..."

While reading through and analyzing the POPs canon, we observed that the teachings therein, again and again, express the teachings of Existentialism regarding the nature of human meaning in the world. This expression is heard throughout the canon. We will focus on three POPs to lend weight to our observation: Lawrence Krauss, Steven Hawking, and Michio Kaku.

Before we begin, let us be clear that all people are free to call themselves whatever they please. We cannot force the POPs to tag themselves as Existentialists. However, when they avow science and scientific method while denying their views are based in any philosophical *a priori,* we say, "If it walks like a duck and quacks like a duck..." Or, as the Japanese say, "By tasting, everyone knows the difference between shit and miso." POPs do indeed quack like Existentialists, and after digesting their books, we are convinced they taste like Existentialists as well.

Now, Lawrence Krauss embodies this Existentialist assertion in his public style and statements as he counsels us to bootstrap a meaningful life in a universe that likely came into existence without a purpose, without human beings at its center. Krauss then looks to the great Absurdist Existentialist philosopher Albert Camus in his seeking for the perfect story to support his belief that we must assert our own meaning in life. "[I] have likened the scientific effort at times to his eternal task of pushing a boulder up a mountain... As Camus imagined, Sisyphus was smiling, and so should we. Our journey, whatever the outcome, provides its own reward."

And finally, there is this ultimate statement of Krauss's existential way in the world, "... such a universe is invigorating. It makes the fact of our existence even more amazing, and it motivates us to draw meaning from our own actions and to make the most of our brief existence in the sun..."

Krauss would find it difficult to ignore this as a philosophical statement. It could stand fast in a Wikipedia article as the definition of the Existentialist goal: to assert meaning in life through one's own independent actions.

As we stated before, the POPs canon is surfeited with Existentialists seeking meaning. Stephen Hawking took a rather ontological approach to the search and comes to an über conclusion on the matter. "It's pretty remarkable to think that our brains... have this wonderful ability to not only perceive reality but to give it meaning too... Meaning can only ever exist within the confines of the human mind... In many ways, this makes us the lords of creation."

And finally, Michio Kaku reveals his Existentialism in the context of discussing the scientific search for the grand unified theory. Kaku acknowledges the search to be motivated by the same force that religious of the past felt: to find our role in the universe, "You have to find meaning for yourself. You have to work for it... there is meaning in the universe but you have to create it for yourself. Otherwise it is a cop out... Meaning lies within yourself."

Thus, it is quite clear, based on the writings of popularizers of science, that they ply their science as Existentialists. We witness this in their declaration of God as irrelevant and their self-assertion of meaning. However, there is one conspicuous paradox that remains. Remember that Existentialist Sartre and Existentialist popularizers alike sense that human beings' existence in the universe is not divinely ordained, that there is "no floor beneath us" and no God to guide us through life. This very sense defines Existentialist nothingness for Sartre, a nothingness that can, and must, be experienced as nausea and anguish. (But according to Nihilology, Sartre's nothingness must be distinguished from Nothing.)

If indeed, POPs view the world through a lens that is ground to the same tolerance as that of the Existentialist, how can we understand their declarations that they actually live in psycho-emotive universes quite at odds? In specific, why are the popularizers of science not nauseous at their nothingness; why are they not full of anguish at being free to assert meaning? Remember that Sartre held as a foundational tenet that nausea and anguish is unavoidable for self-aware human beings. It is the very pith of being alive as an individual.

Sartre himself put forth two quite intriguing and convincing arguments for the lack of nausea and anguish in an individual's life. He held that many people are absorbed within the story of their lives. In other words, the object of consciousness for many in no sense involves their very own self. Now, popularizers of science are by nature focused on the outer world, hence the common phrase "objective science." They do not include themselves, by choice, in their science. This stance is what Sartre called "non-reflective consciousness."

We discussed earlier how humans, according to Existentialist thought, are doomed to be at a distance from their own being. Non-reflective consciousness is an extreme consequence for people who are utterly captivated by the objective world, without their own self in it. Sartre himself employs a long and quite esoteric story of watching a waiter trying in vain to act like a waiter to illustrate his intent here. Let us, for the sake of brevity, be guided by this simple, but explanatory, story borrowed from Paul Vincent Spade.

Let's imagine you are reading a gripping murder mystery. While so engrossed, the object of your consciousness is the story of the murder and the more you become caught up in the story, the more you tend to lose awareness of your environment and your sense of self hood. In particular, in no sense is the object of my consciousness in this case me. We often say that we have "lost ourselves" or are "absorbed" in the mystery's world.

But then, of a sudden, the experience arises in you, "I am really intrigued by this book. The plot is ingeniously concocted." This experience has given birth to the "I" as an integral aspect of your awareness. Suddenly the object of your consciousness is you.

Sartre observed that there is a challenging sense of physical "wrenching" involved in passing from the non-reflective state of consciousness (engrossed in the mystery) to the condition of being a reflective human ("I am intrigued"). It seems the POPs have not been able to maneuver the "wrenching" that is required.

Secondly, all popularizers of science take for granted that they are freely choosing the direction and meaning in their lives, yet according to their published writings and speeches, experience a life without the "haunting" of anguish. Sartre would demand that they are living under the sway of bad faith (*mauvaise foi*). Now this phrase carries much more meaning to English speakers when translated via connotation rather than denotation. In other words, they are under the sway of "self-deception." Sartre would demand that POPs are deceiving themselves in order to avoid the anguish of their life dilemma. In this sense, the popularizers are naive. Sartre goes for the pith of the matter. "The goal of bad faith, as we said, is to put oneself out of reach; it is an escape."

Now, please allow us an additional interpretation of our own regarding this puzzling observation that popularizers of science and Existentialists reside in the same world descriptively and yet live with such drastically divergent human experience? POPs do a great service in bringing once-esoteric science to the literate lay public. In so doing, they receive great laud from the public. We hold that this laud shields them from Sartre's nausea and anguish; laud serves to buoy them up, allowing them to float above the sense of the nothingness below. Furthermore, this popularizing of science transforms the lay public into followers and the scientists into POPstars. Unfortunately, as we now shall discuss, this persona, due to their non-reflective consciousness, their bad faith, is injurious to their scientific inquiry and discovery.

According to our analysis, popularizers of science are playing, according their self-proclamation, the role of übermenschs in today's scientific culture, an exoteric take on the esoteric role Friedrich Nietzsche saw himself playing in late nineteenth century Europe: the *übermensch* of the pre-Einsteinian religious culture.

Recall our discussion before where Nietzsche's Nihilism bespoke the death of God. With God "in the tomb," there was cultural space for certain special humans to bring over-wisdom to the people, to become "lords." Nietzsche envisioned himself as the first of the übermenschs, as witnessed in his immersive identification with Zarathustra. Let us quote again from his sweeping philosophical novel *Thus Spoke Zarathustra*, presented here in stanza format to emphasize the grandeur of his words:

Dead are all gods: now we want the übermensch to live

Man *is* a rope, tied between beast and übermensch—
a rope over an abyss
What is great in man *is* that he is a bridge and not an end
what can be loved in man is that he is an overture and a going under

My friends, a gibe was related to your friend
"Look at Zarathustra! Does he not walk among us as if we were animals?"
But it were better said: "He who has knowledge walks among men as animals."

First, we should note the megalomaniacal superiority that Nietzsche exhibits in his sense of knowing so much more than the commoner, and that his imparting of that knowledge he likens to a human walking among animals. Now Nietzsche was obsessed with ranking humans according to a heuristic scheme of verticality. Commoners among people he termed the *untergang*, the going down as the setting sun. Those exemplars of the masses with a capacity to hear the teachings of the übermensch he called *höheren*, the higher ones. Then there is the *über*, the over, who is a new breed of human being. (Nihilism is Nietzsche's perspective on the life and culture of the *unter*, the below.) All animals rank lower than humans of all types, so the masses, likened to animals, must indeed address the übermensch as "your highness."

We have detected a similar sense of superiority among popularizers of science. They sacrifice their advanced language of mathematics and descend into the argot of the lay public, using stories and images to speak to them at a level they can comprehend at this stage of human evolution. Remember that we sensed the aroma of superiority in the discussion before, as POPs told believers where they were allowed to find God.

Furthermore, in chorus with Nietzsche, the popularizers of science, as we delved into before, boldly state "dead are all gods," that personal gods of religion are at best unnecessary, at worst a dangerous fallacy. POPs also believe, with Nietzsche, that humans, as also discussed earlier, are an "overture," an opening prelude to the future drama that is human evolution from the *unter* to the *höheren* to the *über*, from the under to the higher to the over. Along the way, if we humans are to understand our world correctly, that is empirically, we will become scientists! And POPs are those above us showing us the way.

With all this, it is our contention that popularizer scientists experience themselves in the personae of modern-day, scientific-worldview übermenschs. Where classic scientists from Copernicus to Kepler to Newton to Maxwell lived in eras of religious authority that demanded they strive to reconcile their empirical scientific discoveries with the teachings of the Church, today we live in an era of scientific usurpation that forces the religious to strive toward reconciling their beliefs with the empiricism of science. Indeed, assuming the personae as übermenschs for this era has stationed the popularizers "over" scientific culture, as the messiahs of scientism.

We employ the term "personae" awarely here. Indeed, it must be kept in mind that "persona" derives from the Latin word for a mask worn by characters in a drama. The POPs are acting a part, not unlike Sartre's waiter trying to convince himself through demeanor and voice that he was a waiter. The übermensch of popularizer scientists is a mask of non-reflective consciousness under the sway of *mauvaise foi* or bad faith.

Let us now present the words of three übermenschs of popularizer science: Frank Wilczek, Michio Kaku, and Lawrence Krauss.

As we highlighted earlier, Frank Wilczek believes that the inherent nature of the human being and the inherent nature of the world is one of beauty, and a great discovery of science occurs when the scientist recognizes this covalence of the manifest universe. In his popular book *A Beautiful Question*, Wilczek illustrates this perspective by putting his own wrinkle on a traditional Christian story from the New Testament. He relates how Jesus welcomes the demanding inquisitiveness of his apostle Thomas who is quite taken and excited to discover that reality actually conforms to his deepest hopes. "Doubting Thomas is a hero, and a happy man... Those whose faith is not passive, but engages reality, will receive a... fulfilling blessing in the harmony of belief and experience. Blessed are those who believe what they see."

In reading his whole volume, it becomes quite clear that Wilczek identifies with Thomas, who holds firmly to the heuristic value of faith. Like Thomas, Wilczek engages reality in his scientific endeavors, and upon discovering beauty he is blessed "in the harmony of belief and experience." Wilczek takes on the persona of an apostle of science, an übermensch in the faithful quest for beauty.

Secondly, have a listen to Michio Kaku in a dramatic video lecture titled "The Universe in a Nutshell." As a reminder for all of us as to just how popular and influential YouTube has become in spreading the POPs canon, this video has been viewed by 7.8 million people at the time of writing!

Now, Kaku is a charismatic POPstar full of fascination for the future of humankind. He opens his performance with a specific prediction that, by the year 2100, humans will have "the power of the gods," able to manifest objects from pure thought and fly across the universe in chariots. He is also quite full of himself and envisions his leading role in making this future a normal reality, "I am a physicist... If you want to understand the future, you have to understand physics... "

Kaku then returns to the mid-nineteenth century to highlight his hero James Clerk Maxwell, one of the saints of the lineage of science who codified the mathematical physics of oscillating electric fields and magnetic fields. As Kaku conveniently notes, "Maxwell was a theoretical physicist, just like myself."

Kaku then reveals his übermensch aspirations in quoting a tee shirt styled on Maxwell's equations that was popular in his graduate student days, "In the Beginning God said: 'The four-dimensional divergence of an antisymmetric second rank tensor equals zero.' And there was light, and it was good. And on the seventh day he rested."

We must attempt to convey the atmospherics of this scene in order for the connotations of the tee shirt quote to be appreciated. There are dramatic slides of classic sky/sun/clouds projected in the background reminiscent of scenery from the movie *The Ten Commandments*. There is also choir music adding resonance as Kaku is quoting this tee shirt. We as viewers are being cajoled into hearing Maxwell's equation as holy proclamation; that is, the old story of Genesis is being replaced by the new equations of science.

So, what we watchers see in plain view is that Michio Kaku equates himself with James Clerk Maxwell. He then has Maxwell proclaim the words of God. Thus, Kaku is portraying himself as a replacement for God. This is the very statement of Friedrich Nietzsche: God is dead and I am the übermensch who takes the world into the future.

Kaku goes even farther in his usurpation of God in the future. He reminds us viewers that Maxwell came up with a single mathematical equation that expressed the essential nature of light. Similarly, in seeking a grand unified theory he wants to discover a single equation that will express the nature of all of reality. In his mind this equation will result from the fertile research into string theory, which he believes will reveal the mind of God to be cosmic music resonating through eleven-dimensional hyperspace. "The world I live in, as a theoretical physicist, is not quite the world that you live in. I live in a world that is eleven dimensional."

Michio Kaku predicts that the mind of God will be expressed in a single mathematical equation that he as a physicist will discover. The equation is the score for the essential cosmic music that plays in eleven-dimensional hyperspace, a space in which he happens to reside. Kaku has assumed the persona of übermensch.

Finally, let's take a closer look at the persona of Lawrence Krauss. In the afterword of a recent Krauss bestseller, Richard Dawkins, lead rider in the four horsemen of the new atheism, likens Krauss to Charles Darwin in his impact on the history of science. Recall that Darwin is one of the saints of the POPs lineage, along with Copernicus, Maxwell, Newton and Einstein. The connotation here is: Just as Darwin's evolutionary science undercut the theologians of his day regarding intelligent design, so also is Krauss's scientific cosmogony the knife that cuts the legs off today's theoretical theologians about the creation of the universe.

Now, Krauss does not shrug off this likening to Darwin, but accepts it with a sheepish grin, as witnessed in a talk to Google employees. Yes, it is true that Krauss has received great laud from the popular science world. He was granted the Public Service Award from the National Science Board. This is an honor given only to those who have made substantial contributions to popular understanding of science and engineering in the United States. Also, *Scientific American* has deemed Krauss "a rare scientific public individual."

The dust jacket of the hardcover version of *A Universe from Nothing* states that Krauss is "one of the few prominent scientists today who have actively crossed the chasm between science and popular culture." Krauss indeed has chosen to leap across this chasm. He has chosen to be a figure in popular culture. And, as a result of his great charm and charisma, he has received huge shout-outs from the literate public. Yet this laudation has gone to his head, and he has been coerced into overplaying his persona. (This is not new. Remember Michael Jackson whose moonwalk changed the way people look at motion. A great scientific breakthrough!)

We must insert here a short aside as a means of understanding more fully the persona of Lawrence Krauss. In researching his entire body of writing, video performances, and panel debates, it struck us that he, among the entire society of POPs, most self-confidently asserts the "over" status of science relative to religion and philosophy. This is once again Krauss evolving the role of Darwin.

Krauss goes farther by stating in his haughty manner that the findings of science are much more intriguing than the revelations of religion or imaginings of human philosophizing. "... science *has* changed the playing field, so that these abstract and useless debates [among philosophers] about the nature of nothingness have been replaced by useful, operational efforts to describe how our universe might actually have originated."

In this quiet barrage of slights, we hear Krauss's belittlement of philosophy and religion: abstract and useless, imaginative stories and human concoctions to merely justify the universe. From our point of view, there is really no useful, operational reason for him to flaunt these attitudes. Philosophy, religion, and science have their limits and strengths and should work together toward deeper understanding of our universe. This is the **pattern** approach of Nihilology.

Nonetheless, Krauss piles on. "Indeed, I have challenged several theologians to provide evidence contradicting the premise that theology has made no contribution to knowledge in the past five hundred years, since the dawn of science."

This is Lawrence Krauss jiving us to prove to him that theology has aided science in any way! First, we would point out that Martin Luther's translation of the Bible into the vernacular led to a standardization of the German language that fostered the spread of reading knowledge throughout Europe. In more recent days, Tibetan Buddhists, making use of biofeedback and SPECT technologies, have led the way in our understanding of the effects of meditative states on delta brainwaves.

Second, were the *a priori* of this challenge plummeted, it would reveal the notion that science is of greater status and import than theology, and thus it behooves the theologian to prove their worth to science. In response, we say to all scientists: Provide evidence contradicting our premise that science has made no contribution to theology in the past five thousand years, since the dawn of religion!

With all this context in mind, let us return to our discussion of Krauss's POPs persona, as witnessed in a YouTube video titled "Life, the Universe, and Nothing." Therein he addresses what he, ironically, calls an "incredible conceit" held by human beings. "In biology we seem to think that we are the pinnacle of evolution, that it stops here. Of course, it doesn't stop. Life here on earth continues to evolve. We continue to evolve in our own ways."

So, Krauss likens himself to Darwin, one of the saints of the POPs lineage who brought evolution to the literate culture in its march to overturn the dominant religious worldview, to evolve it to the scientific worldview. Krauss, as the modern-day Darwin, sees himself as the leader of common humans into the future heights of evolution, a prophet who sees the ordinary view of life of the ordinary human as old and mythic and useless. He is bringing down, as from Nietzsche's mountain to the plains of the *unter*, an evolved science of the universe. Krauss assumes the persona of übermensch.

Now, we as Nihilologists have looked closely at the whole extent of the canon of popularizer scientists and we have been driven by its volumes to conclude that this bent toward assuming the persona of übermensch has given birth to an extreme hubris, hubris regarding the POPs themselves and hubris regarding the capacity of rational thought and experimentation to make plain the mysteries of the universe.

Throughout this book we have exposed an ambient urge among POPs to become the supreme explainers of objective reality. Recall the hubristic belief of Lisa Randall, "Nature hides many of its most interesting mysteries. Yet, scientists are arrogant enough to believe we can solve them." Also, as witnessed in the words of Alan Guth and Brian Greene, POPs hold the puffed-up belief that the entirety of reality can be encapsulated in a single unified theory, what Guth calls his "inspiration" and Greene calls "the holy grail of modern physics."

Further, Neil deGrasse Tyson and Brian Greene in their discussion of God of the Gaps express a high form of hubris. "We scientists will eventually leave no gaps in which God can reside!" And finally, we revealed the high talk of Lawrence Krauss who typifies the teachings of religion and philosophy as "useless debates" while his science offers "useful, operational efforts."

The hubris of scientists is not just with our notice come to light, for indeed as early as the seventeenth century Blaise Pascal, the French physicist, theologian, and progenitor of computers, was calling them to task for what he called the "vanity of science." He warned against falling too deeply in love with scientific knowledge as it led the scientist away from thinking about eternal truths, settling instead for the fruits of finite thinking. Of course, we could point to ancient Christian warnings about knowledge, as we discussed earlier regarding the basic flaw in the world. Remember, POPs do not consider themselves heirs of the sin of Adam and so do not believe that seeking rational knowledge about every manifest thing is hubris that will lead their downfall.

We may even hear the warning against rational hubris come from among the popularizers of science, if we are willing to seek widely and ignore the din of the predominant POPs chorus. Listen to the words of our trusted friend the Brazilian physicist Marcelo Gleiser from his book *A Tear at the Edge of Creation* where he advises scientists to break free of the demand for perfect explanation and to move on to the actuality of a scientific worldview that delves into a universe that is imperfect and to accept that knowledge has limits. "…science has its limitations, as do the scientists that do it… We should confess to our confusion and sense of being lost as we confront a Universe that seems to grow more mysterious the more we study it…"

This phase in our critique of the popularizers of science is at heart a call for reflective consciousness on the part of the POPs regarding their role as übermenschs and regarding their hubristic belief in the limitless capacity of rational investigation. It is the perspective of Nihilology that the scientific understanding of the POPs suffers, as we hinted before, when carried out by übermensch personae under the sway of bad faith employing supposedly unbordered rational thought.

Let us call on the writing of the great mid-twentieth century French philosopher, Gabriel Marcel, to help us find a path forward for POPs to get beyond their suffering. As a foreshadowing, we will suggest a new form of research and verification, what we call Apophatic Science.

Now, Marcel is famed in the history of philosophy of science for his ingenious and powerful epistemic differentiation between a scientific problem and a universal mystery. For Marcel, as witnessed in his famed *The Mystery of* Being, a problem is an unknown that can be attacked with the weapon of intellect and reduced to its component aspects toward finding a solution. A problem lies utterly outside the scientist, residing only in objective space. In contrast, listen to Marcel's poesy on the nature of mystery:

> The recognition of mystery is a supremely positive act of mind
> cannot be, strictly speaking, self-conscious
> can grasp itself only through the modes of experience in which its image is reflected
> and which it lights up by being thus reflected in them

[A] mystery is something in which I am myself involved
a sphere where the distinction between what is in me
and what is before me loses its meaning

We have called on the writings of Marcel in this discussion because his "mystery" embodies and echoes much of what we Nihilologists have designated as "Nothing." His statement that "the recognition of mystery is a supremely positive act of mind" is akin to our foundational tenet that Nothing is a declarative verity with a pull to it. This positivity also echoes our profession that the heuristic power of faith serves as a "leaning-into" for the mystery of Nothing. Further, a mystery cannot be "self-conscious," which for Marcel is a declaration that mystery is beyond the reach of intellect. Just so, the Nothing of Nihilology resides over the cliff edge of the conceptual and meta-conceptual in the realm of the super-psyche.

We suggest to the popularizers of science that they heed the epistemic analysis as put forth by Marcel and Nihilology. That the POPs differentiate between a problem and a mystery. That they embrace the challenge of solving problems about the universe **and** embrace the unsolvable mystery of the universe. This embrace of the rational and the extra-rational is indeed the rational thing to do. This embrace will bring about a reflective consciousness regarding POPs übermensch personae, mollify their hubris, and improve their science as well.

Now, we are employing the term "embrace" deliberately here, for it is only in the greeting of the scientist and mystery that the mystery is, as Marcel says, "recognized." In this we hear yet another echo of Nihilology which holds that, in the greeting of the human being as Nothing with the world as Nothing, inperience of Nothing arises and Nihilitive Knowing blooms.

Furthermore, Marcel tells us that mystery is grasped in its reflection, in its image in the human being. Mystery does not reside in the objective alone; in grasping mystery the grasper is integral.

Now, recall our earlier discussion of how POPs as storytellers have insinuated their personhood into the narrative of science, of how the Heisenberg Uncertainty Principle has chiseled a crack in scientific objectivity, and of how the Anthropic Principle has made the human integral to the created world. It appears that POPs have already taken a few tentative steps toward our recommended embrace of mystery.

And finally, recall our extensive discussion before of Apophatic Theology, the *via negativa* or way of denial. The foundational doctrine of the negative way is that it is not possible to encompass the nature of the divine as Nothing via rational thought. We argued that the way of denial is subtle in its ability to differentiate between the nothingness of the manifest world and the Nothing of the divine. In the terminology of Marcel, to differentiate between a scientific problem and a mystery of the universe.

The foundational assumption of Apophatic Theology is that the essential nature of the divine as Nothing is ineffable and so human beings must be satisfied with language that is inept to truly capture the divine as Nothing. In the words of John of Damascus, language cannot convey what the divine as Nothing is, but rather what it is not. Or in the words of Eriugena, we can learn from human language *that* the divine Nothing is, but not *what* the divine Nothing is. Nonetheless, the apophatic thinkers continued to speak, because speaking about the Nothing of the divine was discovered to be an effective way of going beyond the very words uttered.

We recommend that the popularizers of science learn from the apophatics and model a new kind of scientist upon them and a new kind of science upon their wisdom. The new POPs would not declare the death of God, in that they would assert the declarative verity of the divine as Nothing. They would descend from their megalomaniacal *über* to the realm of the *höheren*, the higher intellectuals. The new POPs would evolve from hubris to humble in the recognition that their higher intellects could, with great tenacity and focus, solve the scientific problems of their day. And, at the same time, they would step toward the mysteries of the universe with a heuristic faith that leads them into embrace. Finally, the new POPs would keep on lecturing, keep on debating, in the conviction that they and their listeners may transcend their very words. We recommend POPs transform their profession into Apophatic Science.

Now, it would be fitting, and even a bit satisfying, to end our book at this capstone point in our critique, perhaps with a coda statement in the vein of Stephen Hawking, "Popular science is dead." Or a taunt at Frank Wilczek, "Blessed are those who believe what they cannot think." Or we might mimic Lawrence Krauss, "Popular science is unnecessary—or at least redundant."

But our satisfaction would only tickle us for a moment and would not aid in the evolution of this culture-wide movement to popularize science. With a nobler dedication in mind, we will go one final step in our critique by proposing four specific ways in which Nihilology can make effective contributions to the POPs as they... we... work to solve problems regarding the nature of the universe.

(7) Contributions of Nihilology to Popular Science

Each of the four themes we will explore in this culminating section has a dual intent: first, to deepen the understanding of popularizers of science, that is, to make the POPs better scientists, as we seek together to fathom our reality, and second, to augment the truth value, and illustrate its positive effect on science theorizing, of one essential aspect of Nihilology's teaching on Nothing.

Complementarity suggests that it is incumbent upon popularizers of science to employ multiple languages to create a valid description of the universe. Nothing lies beyond the cliff edge of description. **Instability of Nothingness** calls on perspectives from outside the POPs canon to make a clear differentiation between Nothing and nothingness. **Simulation Hypothesis** argues that the manifest world, *qua* manifest, is a virtual reality that is made real by the ever present Nothing. **String theory** lays out a noteworthy new Nihilological pathway to the discovery of an extra spatial dimension, a material dimension, based on utilizing the human eye as a tool of sensation. Nothing resides in a realm outside of dimension.

Complementarity

Niels Bohr first posited the principle of complementarity in 1927 amidst an atmosphere of grand new notions that gave birth to the paradigm shifting quantum mechanics. It was a period of intense disagreement among the leading physicists of Europe; Albert Einstein, Werner Heisenberg, and Erwin Schrödinger were all in this fray of irreconcilables. This atmosphere of learned conflict turned out to be perfectly appropriate for Bohr, for it stimulated and paralleled his intuition about complementarity of description regarding the quantum world. His principle states that it is often demanded that scientists apply two seemingly opposing explanations to a phenomenon in order to fully describe it. The classic example is the need to typify light sometimes as a wave and sometimes as a particle. Another example is the discovery that a quantum particle's momentum and position cannot be measured at the same time.

Bohr extended his principle to the experimental apparatus in his physics laboratory. He realized that a single experiment could only partially discover the nature of quantum objects. For example, one experiment might reveal energy or position of a quantum object, the so-called particle dynamics, while a second experiment might reveal its wavelike properties. But, in the end, no single experiment could comprehensibly reveal both. And Bohr understood the two or more experiments that are required for a comprehensive explanation to be a complement of another.

Over time and with verifying experimentation, Bohr came to see complementarity in all corners of physics; the theme permeated his later thought. Bohr found, much like us as Nihilologists regarding Nothing, that once he became fully aware of complementarity, he detected its presence throughout the physical universe. This is a trusted sign that

Bohr truly understood the dynamics of complementarity; it became integral to his entire scientific perspective. We might note that, when knighted by the Danish government in 1947, he even incorporated the yin and yang sign into his coat of arms. The Latin motto on Bohr's coat of arms reads *contraria sunt complementa:* contraries are complementary.

Now recall the fascination Frank Wilczek, MIT Nobelist in Physics, carries regarding the yin and yang symbol. This shared fascination with his progenitor lends him an expansive understanding of Bohr's complementarity, for they both interpreted the two swirling fish as equally necessary, but different, aspects of a whole view. (Wilczek quips that Bohr probably had a successful marriage!) Wilczek then goes to the pith of this symbol regarding science. "To do full justice to reality, we must engage it from different perspectives. That is the philosophical principle of complementarity. It is a lesson in humility that quantum theory forces to our attention."

Note here how Wilczek has ballooned Bohr's complementarity, blowing it up into a principle of the successful human marriage and beyond that into a guiding philosophy for all attempts at an explanation of reality! In this expansion, Wilczek has taken a large step toward Nihilology's tenet that knowledge about our reality can only be fathomed in the **pattern** of teachings from canons of religion, philosophy and science. POPs who fail to recognize the pattern, who focus only on the teachings of science in their research, are like the juggler who only sees one ball in the air and ends with the whole juggle tumbling to the floor.

Please allow us a short aside here. Nihilology dares to balloon Bohr's complementarity of knowledge out beyond the quantum realm, beyond the physical world, beyond even Wilczek's complementarity of marriage and philosophy of explanation. For indeed, the pattern of religion, philosophy, and science inculcates the heuristic of faith that allows us to lean over the cliff edge into the realm where Nothing lies. This heuristic allows us to posit Nothing as a declarative with a pull to it. Further, the pattern also points us toward a sort of knowing that is off the cliff edge of explanation itself, what we call Nihilitive Knowing or inperience in that it is a greeting of Nothing by Nothing. As we discussed earlier, and as we will revisit in the third theme of this section, Nothing is ever-present within the manifest world and indeed it is this Nothing that makes the manifest work, like the empty hub of a wheel allows the wheel to turn. With this in mind, we suggest that Bohr's *contraria sunt complementa* applies also to empirical knowing and Nihilitive Knowing. Each is needed to understand a phenomenon. And finally, POPs may need the complementarity of science and Nihilology in order to completely explain all phenomena.

Let us return to our thoughts on the pattern and notice that other popularizers of science have taken small steps toward Nihilology's proposition, breaking from the predominant POPs science-above-all approach.

For example, Albert Einstein, the elder of this movement in culture to popularize science, held staunchly to a view that religion and science have a reciprocal relationship. Max Jammer, the Israeli physicist and historian of physics, was a close colleague of Einstein while at Princeton University where they had many extra-laboratory talks about such matters. In his book *Einstein and Religion: Physics and Theology* he lets us in on those talks.

Jammer cites Einstein's famous proclamation, "Science without religion is lame, religion without science is blind." Harkening back to Aesop's tale of a lame man riding on the shoulders of a blind man, Einstein suggests that science is in need of the "feet" of religion to move about in the various realms of knowledge regarding the universe while religion needs the "eyes" of science to avoid the stumbling stones of naive believing. Aesop stands with us; science and religion need to work together because they both suffer from their own respective disabilities.

Jammer then offers this penetrating insight, "[Einstein] never based his religion on logical inferences from his scientific work." Let us remember that Einstein was culturally Jewish yet was not a believer in a personal God. He experienced a religious feeling regarding the universe to which he reluctantly applied the name pantheism, a feeling that divinity permeates manifestation. Allowing this, we can appreciate the import of Jammer's statement in our discussion of the pattern of Nihilology. Einstein held a free-standing, self-justifying religion in the form of religious feeling. It was not in need of scientific verification, just as the feeling of awe of the popularizers gazing into the night sky reveals a sense of divinity in the manifestation without proof. In the same way, science is free-standing, without need of the imprimatur of religion. Science and religion are two balls in the juggle we refer to as the pattern of knowing.

Sean Carroll, the popular CalTech cosmologist who we critiqued earlier, also has a view about explaining the universe that tracks along a path toward the pattern. In a section of his most well-known book, *Big Picture*, he opines that science's best approach to describing the world is not a single story but an interconnected series of models, each applicable in its own domain of meaning and each to be considered as real. "Our task is to assemble an interlocking set of descriptions, based on some fundamental ideas, that fit together to form a stable planet of belief."

Even though Carroll does indeed take a few steps toward the pattern, those steps are somewhat timid in that he has in mind only models from the science canon, whereas Nihilology suggests the pattern has its source in the canons of science, religion, and philosophy as well.

Now, let us return to Frank Wilczek to see how he comes to his destination on the path toward the pattern. In an early chapter of his book, *A Beautiful Question*, Wilczek shows high regard for Isaac Newton's attempt to approach an explanation of the world from all possible sources of knowledge and via all experiential and empirical tools. He then asserts that his own book is this very same endeavor.

More pointedly, Wilczek spends much of his writing reflecting on the nature of light and human vision. At one juncture he discusses Newton's queries as put forth in his 1706 work *Opticks,* "And here I will add a Query of my own: Is it not unnatural to separate out our understanding of the world into parts that we do not seek to reconcile? It is that Query to which this book, for me, responds."

Wilczek is stating here that he is attempting, in the footsteps of Newton, to seek understanding of the universe via a synthetic view: intellectual, aesthetic, and empirical. It is a search for synthesis employing physics, art, neuroscience, and philosophy to answer the question: Is the universe a thing of beauty?

And finally, Wilczek, in the culminating section of his book, resorts to poetry to incapsulate his philosophy of science. (We have noted elsewhere that POPs have a bent toward the stanza when they need pith in expressing their foundational beliefs.)

I am, and you are, a collection of quarks, gluons, electrons, and photons
I am, and you are, a thinking person

I am, and you are, a material object, subject to the laws of physics
I am, and you are, capable of making choices
I am, and you are, responsible for them

Thus, inspired by the complementarity principle of Niels Bohr, we have shown how this basic discovery of the ineptness of a singular perspective to explain the nature of the world has expressed itself in the writings of many popularizers of science. Each one, from Heisenberg to Einstein to Carroll to Wilczek has understood the need for multiple perspectives in their own peculiar way. We must note that these various peculiarities, in and of themselves, parallel, give witness to and argue for the pattern approach to explaining the universe that Nihilology suggests, a juggle of religion, science, and philosophy.

Furthermore, we have unearthed another inspiration within writings of the popularizers of science that bolsters our argument for Nihilology's pattern in explaining the nature of world. Recall earlier we lauded the research skill of Kurt Gödel, Austrian-American logician of the mid-twentieth century and one of the saints of the POPs lineage. He had a fertile talent for balancing insight and rational analysis in his scientific search, a talent that guided him to employ mathematical reasoning to seek the *a priori* of mathematics itself; he made mathematics self-reflective.

In 1979 Douglas Hofstadter, physicist and professor of cognitive science, brought the work of Kurt Gödel to the eyes of educated lay readers with his book *Gödel, Escher, Bach.* The phenomenon of this work once again reminds us of just how popular the POPs really are. *GEB*, as it is commonly known, ascended to the *New York Times* bestseller list; it also garnered the Pulitzer Prize and National Book Award. When the popularizers of science speak, they introduce a discernible nudge to our view of reality.

In this book, Hofstadter analyzes, more like what urban dictionaries term "riffs on the incompleteness theorem of Gödel, the perceptual art mazes of M.C. Escher and the contrapuntal fugues of J.S. Bach. He concludes that reality is comprised of "strange loops." In the end every system turns back upon itself.

Here is how Gödel formulated his theorem, "To every *w*-consistent recursive class *k* of *formulae* there correspond recursive *class-signs r*, such that neither *v* Gen *r* nor Neg (*v* Gen *r*) belongs to Fig (*k*) (where *v* is the free variable of *r*)."

We see here why POPs leave most mathematical logic out of their canon. Lay intellectuals simply cannot read mathematical language! Note also that this formula is already a translation from German.

Here is Hofstadter's translation into lay-intellectual English, "All consistent axiomatic formulations of number theory include undecipherable propositions."

We should interject here that the incompleteness theorem is not a critique pointing out the incompleteness of mathematics *per se*, any more than relativity theory is about the relativity of epistemic certainty. The incompleteness theorem declares that any first order mathematical language is unable to prove all true propositions. When mathematicians find a true proposition that is unprovable, they are forced to employ a language from outside the first order, such as higher order logic or propositional logic, to prove its truth. The consequences of this are even more profound than those caused by the dilemma of certainty.

In the foreword to the classic book by Ernest Nagel and James Newman, *Gödel's Proof*, Hofstadter warns us that the Gödelian formula is, by its very nature, undermining the hopes of those who believed that mathematical thinking is captured by a single axiomatic system. "[He] thereby forced mathematicians, logicians, and philosophers to explore the mysteries of the newfound chasm irrevocably separating provability and truth."

At some point in plying their trade, mathematicians face the fact that, within their first order system, provability and truth value are trade-offs, not unlike the physicist who faces the fact that the velocity and momentum of a quantum particle cannot be known simultaneously. This "chasm" is the nature of the first order and one must reach above to higher order logic or reach outside for traditional philosophical logic to solve the problem.

Once again, we are bolstered in our notion that multiple perspectives—the pattern of Nihilology— are not just helpful, but an essential requirement in understanding problems about our world.

Gödel's incompleteness theorem also stimulates in us a reverie. Indeed, first order mathematics is a bounded system just as rational thought is itself bounded. Concepts are incomplete in their ability to capture Nothing. Rational thought always finds it border and turns back on itself; it is a "strange loop" as Hofstadter calls it.

We have taken just one deeper breath from our inspiration regarding the incompleteness of mathematics, for indeed mathematics is a language unto itself, the esoteric *lingua franca* of science. And, as we have come to understand, those who speak it are "mathing" the world.

In the early 1950s J. Robert Oppenheimer, the humble physicist and atomic bomb investigator, wrote a book titled *Science and the Common Understanding.* (In retrospect, it was one of the first works by a professional physicist that acknowledged the popularizer movement in the culture.) In an early section, Oppenheimer is describing the findings of the Rutherford alpha particle scattering experiments of 1909, "One could go much farther in describing this discipline even without mathematics; but the words would before long become cumbersome and unfamiliar and almost a misinterpretation of what in mathematical terms can be said with beauty and simplicity."

Niels Bohr, also a physicist in the Manhattan Project, echoes Oppenheimer in his book *Atomic Physics and Human Knowledge.* (Note the parallel book titles.) He upholds the power of pictures and images in the scientific enterprise of explanation in that they propel the scientist beyond thinking to a wider form of creativity. Yet—and here he echoes Oppenheimer true to pitch—pictures and images fail in the description of subatomic activity, for example, measuring momentum and position at the same time. In this case, scientists must resort to (statistical) mathematics to describe both accurately.

Among contemporary POPs, Max Tegmark is the leading voice upholding the inherent power of mathematics to describe reality. Recall from our discussion before, he boldly states that any accurate unified theory of the universe must be wholly created from mathematical equations alone, "...such a complete description must be devoid of any human baggage. This means that it must contain no concepts at all!"

Now we have often pointed out that mathematics serves as the *lingua franca* of science, the common language that allows the entire scientific community of scholars to share the esoterica of their research. As the capstone of long reflection on the pattern of Nihilology, we were granted an a-ha!, an outsight, into the pith of mathematics as language. This outsight occurred while reading an argument, utterly irrelevant to science or mathematics, from Ray L Hart's book *God Being Nothing: Toward a Theogony,* "For the person undertaking this thought-experiment in thinking deity at mind's length... Following the briefest review of the ancient formula *creatio ex nihilo*... the present project was tasked to think the primordial *nihil*."

Note that Hart does not say "thinking about deity," nor "to think about the primordial *nihil*." Rather he says, "thinking deity" and "to think the primordial deity." Hart deliberately chooses his language here to convey the notion that thinking is not merely a descriptive enterprise, rather is a form of creative knowing. Surely, he is not making an assertion of ontic creativity, but he does assert that thinking is an epistemic creator, a making real of the object at hand for the knower. We might say that thinking is akin to fine art painting where a two-dimensional swirl of pigments gives visual access to a world of three spatial dimensions. Further, thinking has its own inherent limitations, much like the disabilities of blind religion and lame science as suggested by Einstein. The limitation, according to Hart, is what he calls "at mind's length." Thinking something into creation implies being at a distance from it.

Our outsight, engendered from reflection on Hart, is that mathematics is also its own form of creative knowing. It defines the knowing of science, and thus we say that scientists "math" their version of reality. Mathematics is the esoteric language of science; it is taught in graduate schools and laboratories around the modern world. Therefore, the science of the popularizers, being sans math, is science in translation. POPs attempt to bring their esoterica to the wider public via translation into reasoned language, stories, and images.

In the same way, religion has its own form of creative knowing in apophatic theology and faith. Apophasis creates reality through negation. Faith is a "leaning into" Nothing that serves as a heuristic for expanded knowledge. Religious "apophase" the world; they "faith" the world. And just like the POPs, religious translate their esoterica into the exoterica of language, stories, and images in order to bring their world to the popular public. Scientists and religious meet in common language. Both need to recognize the esoteric nature of the other's creative knowing and learn from each other in the public domain of language.

Thus, we have different sorts of creative knowing of the world around us. It can be thinked, imagined, spoken, apophased, mathed, and faithed. These are distinct, and complementary, modes of creating reality. All are necessary for a complete understanding of the world. And we suggest that popularizers of science take heed of this fact; thinking, imagination, spoken word, negating, mathematics, and faith must all be taken into consideration and employed in order to have a complete scientific understanding of the world.

We must add a coda that highlights one of the essential understandings of Nihilology and one of its most important contributions to the philosophy of science: Nihilitive Knowing is **not** found among this list of a creative knowings. We intentionally cited Hart's point that thinking is "at mind's length." Thinking something into creation implies being at a distance from it. And the complementary forms of creative knowing all share in this; to apophase, to math, to faith implies a distance between. Now, let us remember that Nothing cannot be **ex**perienced; it can only be a greeting of Nothing by the Nothing of rarefied self. There is, as essential to its nature, no distance between knowing and known object in Nihilitive Knowing. Thus, we termed it **in**perience.

Inperience is not a creative knowing. First, because Nothing isn't a creator; it is only nothingness that brings forth. Second, Nothing lies outside the realm of all description, all characteristics upon which meaning can be hung. The world can be mathed, it can be thinked; it can be faithed, but it cannot be nothinged; it cannot be inperienced. However, Nihilitive knowing or inperience is essential to a functioning world, like the empty hub in the wheel is essential to allowing the wheel to turn. Ever present in the manifest, and essential to thinking, imagining, speaking, mathing or faithing the universe, it is not one of these creative knowings, but it makes all these creative knowings possible.

So, to culminate this theme on complementarity, we suggest that the **pattern** of research from religion, philosophy, and science must lead the POPs. This shift toward pattern would create a new paradigm that we call **apophatic science** wherein POPs boldly seek understanding of the world, yet are free of hubris in the realization that the ultimate nature of the manifest universe suffused by Nothing cannot be known via reason and validation, just as a cascade juggle cannot be performed with only one ball.

To be sure, POPs already approach their profession as scientists and philosophers and religionists, as we established earlier. Though they may not admit, or even be aware of this fact, they benefit from the knowledge of all three fields. They have the pattern of knowing, in no need of our convincing. With this in mind, we suggest that it is a logical next step for POPs to accept the insights of Nihilology and become adherents of apophatic science, one that accepts the cliff edge of rationality, the heuristic nature of faith that aides their leaning over the cliff edge, and the fact of Nothing as a declarative gravitas beyond the cliff edge.

In his classic book *Physics and Philosophy*, Werner Heisenberg, along with his contemporary physicists Niels Bohr and J. Robert Oppenheimer, speak of the transformative effect that quantum theory has wrought on philosophical thinking and on our common cultural understanding of the nature of reality. It is our stance that Nihilology will do the same as the appreciation of its basic tenets are assumed into the theory and practice of POPs science and into the *lingua franca* of literate culture. More pointedly, it is the goal of Nihilology to aid in bringing about an evolution in POPs science, not for some ethical or moral motive, but because applying the teachings of religion and philosophy will make their science more reflective of the actual reality of the world, more productive of guiding theory, and more effective in formulating experimentation. In the final three themes of our book, we shall illustrate with specific examples how this is so.

Instability of nothingness

Throughout the POPs canon we witness the search for Nothing in the manifest universe. For example, physicist Brian Greene, in his book *The Fabric of the Cosmos*, points out that scientists historically had been naive about the nature of empty space which they envisioned as the state of affairs after the removal of every energy and thing. However, with the vision-changing discovery of Nobel scientist Peter Higgs back in the early 1960s, empty space is now taken to be not empty at all.

"Without invoking the spiritual, therefore, we may well closely brush up against the thinking of Henry More [for whom] the usual concept of empty space was meaningless because space is always filled with divine spirit." Greene then reflects that the concept of empty space for scientists is akin to spirit-filled since it is eternally filled with an ocean of Higgs field.

First, let us note that there is a common bent among POPs to deny the reality of divine personage and spirit, then go ahead to employ various teachings emanating from a divine-as-real perspective to deepen their theoretical arguments. In this case, Greene avers that he is not "invoking the spiritual" and then immediately invokes nonetheless, to the betterment of the argument he is making! A more ethical and effective citation of Henry More here would be to ask this question: is it possible that the spirit of More's experience is akin to the Higgs field of science, being differentiated by the mechanism of detection?

On the main point of the status of empty space, most other popularizer physicists second Greene's statement. Alan Lightman, for instance, in a self-revealing article titled "My Own Personal Nothingness" points out that today's physicists employ the term "vacuum" for a region of space exhibiting the lowest possible amount of energy. "But the vacuum cannot be free of fields. The fields necessarily permeate all space... "vacuum" in modern physics is not the void of the ancient Greeks. The void does not exist."

The assertion that the Higgs field permeates all space is important to our discussion. To wit, it is what scientists call a scalar field, that is, its value is not dependent on the reference frame from which it is measured. It is akin to the equal distribution of temperature throughout an enclosed space or the equal distribution of pressure in a liquid. Simply put, the Higgs field will look the same regardless of perspective. And thus, it is a candidate for an absolute frame of reference.

So, popularizer scientists assert that the Higgs field is the utter ground of the world upon which all things stand and from which all things come. Frank Wilczek, in an article titled "Finding Nothing," offers us an easy handle to understand this notion. He suggests that we imagine a race of intelligent fish swimming in water that they experience as their version of empty space. He then concludes, "The big idea I want to convey is simply this: We're like those fish... What our senses perceive as empty space is better understood as a substance, a material."

In our long research into the popularizer literature on Nothing, we have found the book by Harvard mathematician Robert Kaplan *The Nothing That Is: A Natural History of Zero* to be insightful. In a later chapter of the book titled "Is it Out There?" he extols his central theme: there is no zero, no nothing, no void in the manifest world. No nothing in terms of space, time, matter, or energy. When sought through science, Nothing always turns out to be something.

Kaplan muses that Nature abhors a vacuum and so do human beings. Even though zero is knit into our mental fabric, we mistakenly seek its origin outside in physical space. When we assert, in the form of a question, "must there not be a place, numberless places, where there is no matter at all? [We] will hear a dusty reply from our cosmologists...."

In the end most popularizers of science and historians of science agree that seeking Nothing within the realm of the material world indeed is in vain. POPs have done a thorough job of pursuing the Nothing of materiality and found it not. The objective world has been plummeted and, each and every time, in each and every realm of the manifest scanned, Nothing always turns out to be elusive. They conclude quite rightly that there exists no Nothing in the realm of the material.

This conclusion is in accord with the teachings of Nihilology; something lies on a continuum ranging from heavy materiality of stars and planets to utterly light materiality of unstable space and virtual particles. Nothing lies off the ends of the continuum of materiality all together.

We hope our readers will recognize that the POPs' search for Nothing, employing rational theory, outsight, and empirical verification is a cousin paralleling the search for the divine Nothing via apophatic theology as we discussed thoroughly earlier. The more science reaches toward to Nothing, the more something is revealed. This "something" we termed "nothingness." To the ends of the continuum of materiality it is material all the way, turtles all the way down. The apophatics, the master wielders of the negative way, pointed the way: Nothing lies off the end of the continuum of materiality.

Perhaps we should take a moment to reiterate a few other basic notions developed by Nihilology regarding the nature of Nothing. First, gleaned from the teachings of Taoism. Hinduism, and Buddhism as well as the esoteric traditions of Christianity, Judaism, and Islam we learned that Nothing and creation are eternally present. Creation is not the end of Nothing. Nothing is not lost at the creation of the universe. Second, Nothing gives creation what we termed an "isometric balance" that allows created things to function.

Also, regarding the necessary technical language of Nothing, we disciplined ourselves rigorously with a regimen that disallowed any easy humor derived from the multi-valent connotations of the term, nor did we use any intellectually lazy self-contradiction. Without exception, we have employed "Nothing" as a declarative, absent any undercutting or paradoxical negation. With all this we hope to have implanted in the meta-conceptual psyche of our readers a simple sense of Nothing's nature as one of verity and gravity. A Nothing that resides in our readers as true with a pull to it, free of any inherent characteristics or ideational handles.

With all the above tools in hand, let us turn to a critique of the work of cosmologist Lawrence Krauss, a famous POPstar who, among his colleagues, most directly addresses Nothing in his bestselling book *A Universe from Nothing: Why there is Something Rather than Nothing.*

In order to get a sure sense of the meaning of "nothing," readers must pay close attention, for indeed Krauss shifts the ground under our feet a few times over his 190 pages of text. First, we must see that

Krauss uses "nothing" and "nothingness" interchangeably. "Nothing" is employed to contrast with "something" in talking about the stuff of creation. "Nothingness" is called on when an abstract form of the noun is needed to refer to the state of being nothing. However, he does not intend that state to have any deeper connotation. It is simply a necessity of language to bring in the abstract form.

In the main discussion of the book, Krauss proposes three different kinds of nothing in his elucidation of how our universe—something—spontaneously and naturally could have arisen. One of Krauss's nothings is empty space, which he assumes does indeed exist and that the laws of physics also exist. "I suspect that, at the time of Plato and Aquinas, when they pondered why there was something rather than nothing, empty space with nothing in it was probably a good approximation of what they were thinking about."

We find it professionally suspect that Krauss works so hard to differentiate himself from philosophers and theologians, yet as a theoretical physicist he openly avows that his reflections and experiments are essentially akin to the natural philosophers of classical Greece, the later church fathers, and beyond. Does this mean he would have been a philosopher or theologian 2,400 years ago or 800 years ago, but not today?

In any case, since the time of Plato, science has discovered that empty space is full of energy, and this, according to Krauss, allows for our universe to come from nothing, "The average Newtonian gravitational energy of every galaxy in a flat, expanding universe is zero... Our universe has zero total energy and, only thus, can it arise from nothing."

A second sort of nothing is a more radically anti-intuitive proposition on Krauss's part. He holds that the new science of quantum gravity proves that even space is not necessary in the original nothing. Space itself is a natural and spontaneous creation. "The lesson is clear: quantum gravity not only appears to allow universes to be created from nothing—meaning in this case, I emphasize, the absence of space and time—it may require them."

Krauss's third sort of nothing even jettisons the need for pre-existent laws of physics in what he calls the multiverse. "The possibility that our universe is one of a large, even possibly infinite set of distinct and causally separated universes, in each of which any number of fundamental aspects of physical reality may be different... "

Until Krauss, and others, put forth the attack of the multiverse theory, these pre-existent laws have served the religious as a bulwark, an unassailable proof of the existence of God as the eternal *primum movens* who put our "something" in motion. But, according to Krauss, if our universe is not unique, then its laws are also not in need of a prime mover.

Let us point out in passing that yet another sort of nothing is alive in Krauss's book. It will be upon us in two trillion years when the universe has accelerated to such a rate that no body emits light and thus no observations can take place. In the chapter titled "Our Miserable Future" Krauss points to this future nothing and laments that things will slowly get worse for human beings. Galaxies will become invisible as their light degenerates to infrared, microwave, and radio wave. In the book's afterword, Richard Dawkins sums up Krauss's perspective on the future thus, "Finally, and inevitably, the flat universe will further flatten into a nothing that mirrors its beginning... there will be nothing... to see.... Nothing at all. Not even atoms. Nothing."

Please note that Krauss here bases this particular sort of nothing on the fact of it being un-observable, in other words, without human perception. This notion of un-observability will be key in the final theme of this section wherein we discuss the extra dimensions demanded by string theory.

So, to conclude this phase of our critique, Lawrence Krauss is in harmony with the POPs canon regarding the foundational tenet that all reality is material. Materiality is not a continuum with Nothing greeting one off the ends, as Nihilology has argued. However, Krauss takes quite a different tack into the wind of language. Whereas the resounding voice of the canon states that Nothing does not exist, only something, Krauss maintains that Nothing and something both exist and they are both material in nature. Hear his poetics on the matter:

For, surely, 'nothing' is every bit as physical as 'something'
It behooves us to understand precisely the physical nature of both these quantities
And without science, any definition is just words

When I use the word nothing
I actually don't mean nothing
I mean *nothing*, i.e., it has weight.

So, what is Krauss's motivation, on the one hand to empirically verify the nature of reality in accord with most POPs, while on the other hand to put forth a language quite at odds with the majority, a language more cumbersome and no more effective than the majority language?

As we read Krauss's work, we noticed throughout that he seems to carry a small dagger in his jacket whenever out in public, just in case a theologian or philosopher comes by who he deems in need of a poke. He seems to bear a chip on his intellectual shoulder that expresses itself as a zealous drive to overturn the worldview of God-centered religions and western philosophies. It is this intellectual chip that explains his crooked choice of terminology. A close analysis of the bold words displayed on the cover of his book reveals the truth of our perspective here.

Krauss's lead to the intellectual lay audience whom he aspires to convince includes a title that harkens to Christian mythic doctrine and a subtitle that is a classic question drawn from philosophy.

We discussed earlier the argument between Gnostic Christians and the culturally dominant Latin church over whether God brought the universe into manifestation from an eternal undifferentiated stuff, like a potter giving form to clay, or that he created the world from Nothing (*ex nihilo*). With the defeat of the Gnostics the *ex nihilo* side of the argument become canonized doctrine.

Now Krauss's zealous need is to prove that the world indeed came to be out of Nothing but that the act of creation did not require a creator God. In order to do so, he has to alter the age-old meaning of Nothing, bending it to mean a type of something. Just to be clear, in the Krauss lexicon, "nothing" denotes the Higgs field, described above as the utter ground of the world upon which all things stand and from which all things come. Whereas "something" denotes the stuff and energies born out of the Higgs field at the moment of the Big Bang. With this strategy, he forcefully argues that the world comes to be "naturally," making God irrelevant to the deed, yet he falls into the silly position of upholding the defeated Gnostic tenet that the world was created from the modern-day version of clay, the Higgs field.

Thus, the actual title of Krauss's book is *A Universe from Something*!

Further, the subtitle of the book evokes a question taken up throughout the history of human thought. This particular form of the question has been asked by many philosophers in the West: Aristotle, Leibniz, Heidegger, Russell. "Why is there something rather than nothing?" These philosophers were aghast at the fact that the universe of stars and thoughts **is...** when it seems obvious that Nothing could equally have been the eternal state.

First, let us note Krauss's attitude toward this question as stated in the preface to his book. First, he points out that, since the time of Christ, the question "Why is there something rather than nothing?" has gained life as a challenge to the proposition that our universe came into being without design, intent, or purpose. "While this is usually framed as a philosophical or religious question, it is first and foremost a question about the natural world, and so the appropriate place to try to resolve it, first and foremost, is with science."

Krauss is presumptuous here. It is his right to argue that his science is the proper avenue to answer a question, but he is quite out of line telling readers that the question is not really, and never has been, the human question of meaning that they universally experience themselves asking.

Regarding the content of the question, Krauss holds that popularizers of science do not ask "Why?" questions in that this would imply an investigation of purpose, and since Krauss holds to the Existentialist position that life is inherently without pre or extra-human meaning, POPs really intend "How?" when they say "Why?" Thus, the actual subtitle devolves a bit into, "How there is something rather than nothing."

Now, we have quoted Krauss plainly stating Nothing and something are both material. Therefore, his Nothing is not differentiated from something as a qualitative or ontological difference, rather their difference is an affair of degree of materiality. Remember our discussion earlier about the rarefying principle, shared by eastern and western religious traditions with science. It states that the more we take away from an existent thing the more we understand that basic reality of the thing. Nothing, the clay of the Higgs field, is the rarest form of material while something is the manifest heaviest form of material. Thus, the actual subtitle devolves even lower, "How there is heavy materiality rather than rare materiality."

The nadir of the actual subtitle is plunged when we recall that nothing of the Higgs field continues to be present after the Big Bang creation of something. The Higgs field is still present, even now underlying our manifest world. Thus, the actual subtitle devolves to its nadir as, "How there is heavy materiality **and** rare materiality."

Thus, the actual book cover lead is the rather unmanageable: *A Universe from Something: How there is heavy materiality* ***and*** *rare materiality.*

It is our analysis that Krauss clings to his language of something and nothing in attempt to cleanse the minds of his lay intellectual readers of the belief that God created the world out of Nothing. We hold that his arguments partake of shenanigans, and his work could be much more convincing to his audience were he to drop his zealous chip and join the majority of his popularizer colleagues in the more simple, agile, and effective language: there is no Nothing in the material world.

Now to one take final step, essentially all POPs physicists and cosmologists have concluded that the universe arose naturally out of the Higgs field or its concomitant base referred to as the quantum vacuum. They errantly refer to these as "nothing" instead of the more accurate "something," or the more effective "nothingness" as argued by Nihilology, so it may be confusing to our readers to cite the popularizer's statements.

But readers who have stuck with us this far are used to their mis-usage by now. Lawrence Krauss says, "These quantum fluctuations imply something essential about the quantum world: nothing always produces something, if only for an instant." Frank Wilczek says, "The answer to the ancient question 'Why is there something rather than nothing?' would be that nothing is unstable." K.C Cole says, "Potential, as it turns out, is one of the most impressive properties of nothing."

This conclusion on the part of the POPs regarding the creation of the universe, precisely tracks the perspective of the religio-philosophical school Samkhya, as we discussed before in Section #3. The Samkhyans were the masters of observation. They meditated with their eyes open. They scrutinized the world and made, in light of modern scientific cosmology, quite prescient conclusions.

The world in Samkhya is named *Prakriti,* "the going forth." *Prakriti* is said to be utterly constituted of energetic qualities called *gunas.* They are three in number: lightness, heaviness, and movement. The *gunas* provide an analysis of the basic makeup of the world. And, germane to our discussion, when the world dissolves into the unmanifest state, the qualities that make up *Prakriti's* totality continue to exist.

Calling upon the creation language the POPs' *Prakriti* in her unmanifest state *(avyakta),* inherently comprised of unstable productive energies (*gunas*), causes the world to manifest *(vyakta)* out of itself. The world "goes forth" as a something manifested from a pre-existent nothingness.

It is our contention that the perspective of Samkhya exactly parallels the scientific perspective of an inherently unstable quantum vacuum naturally bringing forth creation. Again, we hold that the Samkhya teachers were good quantum mechanics!

Indeed, it seems to be the case that the Samkhya position on the inherent instability of nothingness is point for point consistent with quantum mechanics. And yet, Samkhyans found their position to be utterly consistent with, and inclusive of, the reality of Nothing (*Purusha*) that lies outside the realm of the material. POPs would do well to pay heed to the Samkhya masters.

The popularizers of science are to be congratulated on their tenacious search for Nothing in the material world. Yet Nothing has not been discovered therein. Please allow us to make a suggestion for future research. Remember that the rarefying principle, as practiced by the religious, has its fulfillment in the rarefying of the world and the rarefying of the self. The goal the POPs seek lies in researching the reality of the world in conjunction with just as deep an investigation into the reality of the researcher. Popular science needs to turn toward a search for Nothing in the scientist!

Our call here for a more expansive realm of legitimate research is not at all outlandish nor idealistic. Recall our earlier discussion regarding how the pivot toward including the personhood of the scientist has already begun for the POPs in their recognition of the Uncertainty Principle and the Anthropic Principle. POPs already take seriously the basic notion that the conscious person is integral to the whole of the manifest reality.

POPs could, of course, take on the disciplines of Jewish Kabbalah, Christian esoterism, or Zen Buddhism as we discussed earlier. They could train themselves toward inperience. It is only in the greeting the Nothing of self with the Nothing of the world or divine that the knowledge of Nothing blooms. We call this inperience Nihilitive Knowing.

Now this would indeed be an outlandish and idealistic request. Recall in our earlier analysis of Existentialism, we suggested that the POPs generally suffer from non-reflective consciousness. Thus, it would be ineffective to ask them to expand their experience to the realm of inperience... though this would enlighten them as to their objective experience actually being intersubjective in nature. Inperience would also rid the POPs of all hubris: there is no übermensch in the absence of "human."

So, what sort of effective discipline can we recommend? Taking into consideration all the evidence Nihilology has garnered from religion, science, and philosophy that supports Nothing as a fact of reality, we **can** encourage belief in Nothing. We have shown earlier how faith serves as a heuristic that allows "leaning into," a leaning over the cliff edge of materiality and concept, allowing a peering into realm of Nothing.

This is not an outlandish and idealistic recommendation. Remember that POPs already take advantage of this heuristic of faith in three different forms of belief. 1. The universe has an essence. 2. The universe is of a unified nature. 3. Beauty works as a guide to discovery.

Belief in Nothing would transform the *a priori* of POPs research, for instance, the search for dark matter out at the end of the continuum of materiality, the search for dark energy out at the end of the continuum of forces, or the search for quantum entanglement out at the end of the continuum of space. Each new theory or experiment would begin with the taken-on-faith that Nothing is a declarative with gravitas waiting in a realm just off the ends of those continua. Thus, each would spring from a wider platform and head in new directions.

We have often used the phrases "a declarative with gravitas" or "a declarative with a pull to it." We intended that human beings are drawn to the reality of Nothing. We have witnessed the effect of this gravitas on the POPs who have zealously sought Nothing in their research. Now, we would like to suggest that "gravitas" has yet another meaning.

Science tells us that dark matter is detected in part by its effects on the total gravitational force of the universe. We contend that Nothing has an effect on the manifest universe in a way equivalent to gravity. Recall in the writings of Samkhya that the world takes on a specific form at the behest of the very presence of Nothing. Recall also the oft-referred-to Taoist notion of the empty hub allowing the wheel to roll. Similarly, Nothing has the effect of giving a profound nudge to the manifest world. It has a function in the world even though this function is acausal and non-creative.

As we shall see in the next theme on the Simulation Hypothesis, the manifest world *qua* manifest is a virtual reality. Nothing functions to make the manifest real.

Simulation Hypothesis

In 2003, Nick Bostrom, founder of the Oxford Martin Programme on the Impacts of Future Technology, set off an IED in the roadway traveled by popularizers of science. He capped a trilemma of rather loosely established propositions with this boom! "We are almost certainly living in a computer simulation." Since then, POPs have been unable to look away from the bomb crater. With their accustomed tools they have been unable to repair the pavement; they have failed to solve the problems presented by Bostrom's proposition. Fortunately, with the tools of Nihilology the crater can be fixed.

Now, we must establish that Bostrom is not a fringe thinker. He is an Oxford University professor, a futurist, whose thought is influential in the POPS canon. For instance, his 2014 book *Superintelligence: Paths, Dangers, Strategies* resided on the *New York Times* bestseller list. His warnings regarding the ethical challenges of AI have been taken up by Bill Gates and Elon Musk. Once again, we recall the extreme influence popularizers of science have upon the literate worldview.

So, simply put, Bostrom posits an indeterminate future when a post-human species will evolve with superior intellect and technical power. It will create an all-encompassing computer simulation of the cosmos in an earlier era. He calls this an ancestor simulation. We present-day humans are living in this virtual reality. In an article titled "Are You Living in a Computer Simulation?" Bostrom states, "While the world we see is in some sense 'real,' it is not located at the fundamental level of reality."

Before highlighting the widespread fascination among popularizers of science with the simulation hypothesis, it is incumbent upon us to point out two poignant difficulties with the notion. First, Bostrom holds *a priori* that future post-humans will have the desire to create ancestor simulations, based on his observation that "there are certainly many humans [today] who would like to run ancestor-simulations if they could afford to do so."

Where does he get this idea? Certainly, there have been no scientific research studies that uphold such a desire among people today. We can only conclude this is a personal belief on Bostrom's part. And further, such a desire on the part of post-humans clearly constitutes a purpose for creation of our virtual world. And, as we discussed earlier, science avers loudly that it is not in the profession of asking "why," acting only as the purveyors of "how."

Secondly, Bostrom tells of a colleague who reported seeing pixels when looking out at the objective world, especially in the low lighting of dawn or dusk. The colleague has no discernible eye disease nor vision difficulties of any kind; he considers his pixelated vision to be a positive extra-sensitivity rather than a pathology. He brings up his experience because computer generated images are composed of pixels and thus his vision of the daily world as pixelated bolsters Bostrom's proposition that our world is indeed a computer-generated simulation. Oddly enough, Bostrom dismisses his colleague without reflection for he holds that the post-human programmers would be so super intelligent that they would reprogram anyone of us who experiences a glitch in the simulation. In this he demands that there is no way humans could detect the virtual nature of the simulation. From our vantage point, this is a rather ironic position for a scientist in that he is putting forward an argument that cannot be falsified... or verified! And, as we show at the conclusion of this theme, it is a position that ignores the pattern of knowledge about the world as held by Nihilology.

Now, even though popularizers of science have thus far shown no aptitude for solving the challenge of Bostrom's Simulation Hypothesis, they have continued to be fascinated with it. For instance, Stephen Hawking, in a YouTube video titled "Stephen Hawking Talks About Reality," stated, "The world around you is an elaborate fabrication of some unknown superior intelligence." And Neil deGrasse Tyson has posited that it is likely the universe we live in is a computer simulation. Graham Templeton sums up Tyson's perspective, "... we have to assume that on an infinite timeline some species, somewhere, *will* simulate the universe. And if the universe *will* be perfectly...simulated at some point, then we have to examine the possibility that we live inside such a universe."

Tyson is so taken with the Simulation Hypothesis that he convened an entire session of the renowned Isaac Asimov Memorial Debate to reflect on the problem. Sitting on the panel were many POPstars: Max Tegmark, Sylvester James Gates, Lisa Randall, and David Chalmers. Just in passing, we might note that Chalmers even applies statistics to the question, proposing that there is a 20% chance that our world is a simulation.

The actual content flowing from this fascination of the popularizers of science with a posited simulated quality of our world makes it clear once again that POPs are taking on roles in literate culture that have traditionally been played by religions. It also reveals that they are, with their accustomed tools, out of their depth in this analysis.

First, this whole analysis by the popularizers of science regarding the Simulation Hypothesis turns the query of classic western philosophy of mind on its head. From Pyrrho to René Descartes to John Locke it has been asked, "How do we know the world is real?" The *a priori* for philosophers in the western tradition has been that the world **is real**, yet how can we know for sure it is real. The Simulation Hypothesis has a flipped *a priori:* the world **is a simulation**, yet how can we know for sure it is virtual. POPs express their fascination by attempting to find the glitches in the computer program.

Recall the discussion earlier about how popularizers of science are usurping, or are unawares inheriting, many of the traditional roles of religion in culture. Ever since 1927, with the vagaries of particle measurement introduced by Heisenberg and later with the discovery of the quantum world and its mysterious entanglement between particles at a distance and its participatory creativity that results from simple observation, POPS have cajoled us into believing that our everyday experience of the manifest world cannot be trusted. As examples, we noted Frank Wilczek and Max Tegmark's invitation to join in their mission of seeing beyond the daily sensate experience of the world.

And now with the Simulation Hypothesis they are warning us that the manifest world itself is not real. With this alert, today's scientists, physicists and cosmologists, we suggest, are playing an important cultural role that has traditionally been played by religious leaders, typically in the East, who have been fascinated with analysis of the manifest world and why it is not utterly real. In this, POPs are assuming the stance of religious of the East who begin with the experience of the world as unreal, as *a priori*, and seek reality from a basis in this experience.

Now, were the POPs to do their research based on the pattern of knowledge about the world as witnessed in science, religion, and philosophy, they would understand that there exists among eastern traditions a sophisticated, centuries-old analysis of the illusory or virtual nature of the manifest world.

As we elaborated before, Buddhists of the Mahayana and Zen schools hold that all created things are empty (*shunya*) of realness as independent entities. (They find their reality only in their context of Nothing.) Taoists hold that the manifest world is unable to function *qua* manifest. (It only works because of the permeation of Nothing). Samkhya envisions a world where neither the objective sphere of things nor the subjective sphere of mind and emotion are the reality of the human being. (Only the aware Nothing is the real essence of humans.) However, it is in the teachings of the Hindu school of Vedanta that we discovered the most potent analysis of the virtual world... and in a language that Bostrom and the popularizers of science will find familiar.

Now, Vedanta holds that the manifest world is a superimposition (*avivarta)* upon the absolute without qualities (*Nirguna Brahman)*. The root term here comes from Sanskrit *vrt*, to roll. So, *a-vivarta* implies that the world has never rolled out at all. Indeed, in Vedanta, the world is not a created something! Also, *Brahman* is not a personal deity at all. The Sansksrit word finds its roots in "expansion" and implies here that *Brahman* is infinite expansion, infinity. This infinity is what Nihilology refers to as Nothing and the manifest world is a dreamlike covering-over of Nothing. Recall this pithy verse by Shankara, the main spokesman of Vedanta, "As dream-built lands and times... are all unreal just so here in waking is this world... in as much as all this world... [is] unreal."

The Vedantists call this dream world an illusion (*maya*). It is equivalent to what popularizers of science call virtual reality. VR simulates a three-dimensional space that a person becomes immersed to the point of being utterly convinced that it is genuine. The Vedantist teaches that it is exactly the same with the ostensibly manifest world. With life-long immersion, human beings come to believe in its reality, when in fact it is almost real. Does this not echo Nick Bostrom's Simulation Hypothesis?

To go one step farther, Vedanta posits that illusion misleads us into believing the world has ontological status separate from Nothing (*Brahman*). The world only seems to be differentiated from Nothing. This seeming quality is by force of what Vedanta calls human ignorance: *avidya*, literally "not knowing." Ignorance has us superimpose concepts and names and qualities on Nothing, thus presenting us with an empirical world that we take as real. Bostrom and the POPs should take heed of the Vedantists, not only in imbibing that they are ignorant, but in taking the Vedantist challenge to overcome their ignorance and investigate the possibility of knowing Nothing as the basis of the reality of the manifest world. When the popularizers rid themselves of ignorance, the world will appear as it is: real only with the presence of Nothing. Nothing makes the world real.

Here we'll highlight one popularizer of science who has taken significant steps toward this perspective of Nihilology regarding "not-knowing." Recall that the Hindu school Samkhya calls the manifest world *Prakriti,* "the going forth." It is inherently comprised of unstable productive energies (*gunas*) that cause the world to manifest out of itself. The world "goes forth" as manifested from a pre-existent nothingness. Lying off the end of the continuum of the manifest material world is *Purusha,* the witnessing Nothing that constitutes the essence of the human being.

Now Samkhya avows perhaps the clearest teleological tenet in all of world religions and philosophies: The purpose of the manifest world is for humans to see it. The manifest world allows us to witness exactly what we need toward freeing ourselves from ignorant false identification with a virtual self, a virtual self much like our avatar in the space of a simulation game or the world we live in according to the Simulation Hypothesis. Being an aware witness to the manifest gives birth to comprehending its nature and our nature. Recall this verse from the root text of Samkhya:

> The relationship of those two, like the blind and the lame
> is for the purpose of humans witnessing [the world]
> and for their discrimination
> Creation comes forth out of that relationship

Now, we have cited various times earlier the British physicist at Arizona State University, Paul Davies. He is Director of BEYOND: Center for Fundamental Concepts in Physics and is popular in his role as interviewee on YouTube, such as this one titled "Big Pictures of God." Read his words, here presented in stanza format to emphasize their poetic quality, and note their kinship with Samkhya:

I think we should get away from the notion
that the universe has meaning and purpose
because it is imposed on it by some external being
In the realm of human affairs we see that meaning and purpose
emerges from nature as a part of evolution
In the same way, meaning and purpose can emerge in the cosmos as well

Many scientists say that life and mind
are trivial embellishments in the cosmic story
but I have always believed that life and mind are of cosmic significance
Human beings are more than just observers
We don't just watch the show, we understand it

Something which is the most complex system we know, the human brain
touches, connects with the most fundamental rules on which the universe runs
The evolution of life and then mind that connects into the deep workings
of the cosmos suggests it has cosmic significance
Ultimately, the purpose of the universe includes its self-comprehension
The universe has somehow engineered... its self-comprehension

It certainly is apparent for all our readers that this popularizer of science stands alongside this eastern religion in avowing at least three important aspects of the world's nature: 1) Meaning and purpose are not imposed by a divine being. 2) Meaning and purpose are emergent aspects of life. 3) Human beings are essential to the manifestation as witnesses and comprehenders of the world.

We must, nonetheless, acknowledge that Davies has merely taken a few steps toward this worldview that has informed, in its own way, the teachings of Nihilology. For Davies proposes that the rational thoughts of the human mind are capable of comprehending the nature of manifest life and human life therein. Whereas, as we have argued extensively before, we must be awake in a realm beyond concepts in order to greet our human nature as, what Samkhya calls, the awake Nothing. This being said, Davies has made an essential turn toward Nihilitive Knowing, a turn that the POPs majority has yet to make.

So, as we concluded before, the manifest world is made real by the presence of Nothing, and with our discussion of Samkhya, we can now add that the reality of the human in the world will be experienced with by overcoming ignorance through aware witnessing of the world, leading to what Nihilology has termed inperience or Nihilitive Knowing.

Remember our earlier discussion of how POPs bring science to the literate public by translating from the language of mathematics to that of words, stories, and images. In the same way, religions and philosophies translate their meta-conceptual knowing into words, stories, and images to speak to those of the literate culture regarding the virtual reality of the world and the human being. To be sure, each religio-philosophical tradition has its unique mode of translating its language. For example, as we elaborated in Section #3, Vedanta speaks of the virtual reality of the manifest world. Early Buddhism speaks of the unreality of the individual personality, and later Buddhism expands this unreality to the individual things in the world. Taoism speaks of the yin and yang of the world being unable to function without the context of the ever-present *wu*. The esoteric traditions of Judaism, Christianity, and

Islam speak of the lack of reality of a divine with characteristics. What unites all these translations is the deep belief that the true nature of the world, or the individual in the world, is Nothing. And when one includes Nothing in one's view, the world and the individual in the world is wholly changed. The world finds itself to be real.

Yet, how can we understand this Nothing, a declarative with a pull to it, with neither conceptual handles nor capacity for causation, to have any effect on the manifest at all? The theological canons of the world are teeming with simple, enlightening tropes: the empty hub that allows the wheel to go round, the unseen circle that determines the swim of the yin and yang fish, the influx of *Ayin* that defines the identity of the person, and so on. Here are two analogies, one from the realm of mathematics and the other from that of neurology to aid POPs in reflecting on this issue:

Recall that, when zero first made its debut in ancient forms of calculation, it was held to be a void, an empty space into which a digit could be slotted. It was an unknown quantity which mathematicians now reserve for variables with "X." It took over a thousand years for zero to be recognized as a number in its own right, however it remains a number without quantity. Rather it serves to move all other numbers to their proper place, so that **their** proper quantity is determined.

We point to the writings of Erling Kagge, famed as a member of the first expedition to reach the North Pole unsupported. That trek provided him with outsights as to the nulling of sound. In his book *Silence in the Age of Noise,* he speaks specifically of the effect of *cesura* in music, which is a non-tempo absence of sound. He calls it an "adventure" listening to Beethoven, "da,da,da,daa... but it is the *cesura,* the pauses between—the silence between the noise of the instruments—which are my favourite. That's when I am stirred awake."

Neurological research, as for instance in Valorie Salimpoor's eruditely titled article "Anatomically Distinct Dopamine Release During Anticipation and Experience of Peak Emotion to Music," conclusively shows how the mind registers and processes sound. During interludes, silences and quiet between notes the greatest amount of neural activity in the brain occurs. This intense brain work generates the peak emotions of music. Nothing, indeed, can have a great effect.

With all this we can see how the eastern canon has much to offer the popularizers of science in their investigation into the virtual reality of the manifest world: Humans un-knowing contribute to a virtual world being experienced as real. We people in a virtual world are convincingly akin to the avatars in a simulation game. The most valid *a priori* is that the manifest world can be experienced as a virtual reality. By witnessing the manifest world human un-knowing is overcome.

Popularizers of science would do well to investigate the entire pattern of knowledge, as proposed by Nihilology, regarding the virtuality of the manifest world, for their limited analytical tools of rational argument, experimentation, and verification have shown themselves inept to repair the crater of Bostrom's bombshell challenge of his Simulation Hypothesis. Let us now elaborate on one unalloyed failure on the part of the POPs that serves to point out immaturity in their newborn role as messengers of a simulated world.

First, heed Bostrom discussing certain ramifications of his Simulation Hypothesis regarding what he terms "loose analogies with religious conceptions of the world." He states, without scientific shame, that the posthumans who are running the simulation are akin to gods in relation to us humans who play out our lives within the dimensions of the simulation. "The posthumans created the world we see;... they are "omnipotent" in the sense that they can interfere in the workings of our world even in ways that violate its physical laws; and they are "omniscient" in the sense that they can monitor everything that happens."

In order to go farther with understanding this issue, we must introduce Sylvester James Gates. He is a professor of physics at Brown University who is well-known among POPs for his string theory research. It is interesting to note that he did not announce himself to the popular literate culture via books; he gained his fame through the video series *Superstring Theory: The DNA of Reality*, put out by the Teaching Company in 2006. It consisted of 24 half-hour lectures intended to make plain the esoterica of unification theory for his viewing audience of laypeople.

Now, Gates was a panel member at the Neil deGrasse Tyson's Asimov debate mentioned before. And he holds that there is much more in the Simulation Hypothesis than loose analogies with religion:

> Tyson, "Jim, is there any functional difference at all
> between admitting that we live in a computer simulation
> and saying that that is basically a secular God?"
>
> Gates, "For non-scientists, the acceptance of the Simulation
> Hypothesis
> as an accurate view of our universe, is equivalent, I believe,
> to the notion of a deity. I don't understand how you can make that
> distinction."

Gates boldly takes the quotation marks off Bostrom's proposition: the post human programmers of our virtual world are not "deities," they are deities! The voices of Bostrom and Gates are echoed by many others. They give witness to an attempt to establish a source of the simulation, an ürprogrammer who generates the virtual reality of our manifest world. This attempt is the POPs variant of the classic belief in a prime mover for creation of the natural universe.

Recall our earlier discussions of Gordon Allport's foundational work in the 1950s that introduced us to notion of religious maturation paralleling, though not necessarily coinciding with, personal maturation. Immature religious are focused on magical thinking and self-reward. Theirs is a faith in mythic truth and orthopraxy: Christians in the pew every Sunay, Muslims who perform *salah* five times a day, and Jews who light the candles for *Shabbat.* This puerile religion is admirable and describes the considerable majority of practitioners in the world's religions. Nonetheless, there are a few religious people who mature into believers of doubt who are called to new understanding beyond the bounds of current understanding, still fewer who grow toward the sophisticated *apophasis* of theologians and Nihilitive knowers.

It is among these congregations of every Friday, Saturday, or Sunday religious people that lives the unquestioned prime mover: Allah, Yaheh, or God. To be sure, *primum movens* was first contemplated by Plato and Aristotle and later by St. Thomas Aquinas as the "unmoved mover" that moves other things but is not itself moved by any prior action. However, this sophisticated concept is not the prime mover who lives among our congregations based in myth and orthopraxy. He is, rather, a personal divine being who resides outside the manifest world and who gives the manifest world its original push.

Now popularizers of science, from Krauss to Hawking to Kaku to Randall to Green have vehemently denied the need for a prime mover in the process of bringing forth the manifest natural world ("God is at best irrelevant.") At the same time the POPs argue that, if the world is a simulation, there must be a simulator standing above and beyond the simulation. It seems this is hypocrisy. In a virtual world, the POPs abandon their root belief in naturalist self-emergence, a move that should be anathema. Why does their analysis require such a prime mover in the case of the creation of a virtual world? Is this not an unreflective throwback to puerile religion? The notion that only a future post-human society has the capacity to bring forth a simulated universe is clearly a replacement of a mythic God with a smart people.

Our discussion above points to sophisticated analyses in the eastern canon as to how a virtual world can be created without the input of an outside divine being. It appears to us that Nick Bostrom and the group of fascinated POPs have abandoned their own principles of science, in a panicked regression to the puerile in order to explain a phenomenon that they are flailing to understand.

In fact, POPs have no need of seeking a prime mover in the case of the Simulation Hypothesis any more than in any other scenario of cosmology. They do fine with a naturalistic creation if the world is real. That the world is virtually real does not alter that creative process or ground. The only change is that now the POPs are being challenged to plummet the reality of that natural virtual creation. Nihilology can enhance the understanding of scientists regarding this Simulation Hypothesis, offering a cosmology where a personal unmoved mover God is not needed. Were they to follow the lead of Nihilology, they would also discover that the virtual natural creation is only made real with the presence of Nothing.

Extra Dimensions of String Theory

Earlier, we put forth a heuristic statement, "Scientific discovery, at various epochs of history, has been led by empirical experimentation, by mathematics, or by theory. When led by theory, science is methodologically indistinguishable from philosophy." It guided us in arguing that popularizers of science are philosophers, doing philosophy regardless of their vocal protest. This heuristic is a handy tool once again as we now analyze string theory and propose how Nihilology can aid POPs in their efforts to understand how string theory can, in fact, describe our manifest physical world.

Now string theory is a child of mathematics, and mathematics demands that the other two components of scientific discovery catch up in order to bolster its claims through logical theorizing and empirical verification. This demand is especially urgent, more like a military order, regarding the question of spatial dimensions. For, it turns out the mathematics of string theory requires the existence of at least six extra spatial dimensions in order to solve the overarching problem that motivates string theory, i.e., to unify the realm of the super small, as described by quantum mechanics, and the realm of the super large and fast as described by general relativity. (We elaborated on this earlier in our discussion of the search for a unified theory or theory of everything.)

Unfortunately, POPs have been utterly unable to comply with this order because, **in their eyes**, no viable empirical evidence has been discovered that these six spatial dimensions have any reality outside of mathematics. However, Nihilology has a different set of eyes. We hold that popularizers of science have blinded themselves by relying on their "non-observability" of the extra dimensions. This non-observability has also misled theory's attempt to catch up with mathematics as well. As we go forward in elaborating this final theme of our book, we will argue, from the viewpoint of Nihilology, it is in the demand for the nano-smallness of the extra dimensions of string theory that POPs have been led astray in their research.

In this final theme, we will provide a new path for empirical research into string theory's missing dimensions. In order to do so, we will return to our earlier discussion of what we called the Rarefying Principle. From this basis, we will develop a new model that we tag "Four-Staged Kenosis" that will bring certain of these missing dimensions into plain view.

Now, it was the "Finnish Einstein" Gunnar Nordström who, back in 1914, introduced the world of science—from the mouth of a scientist, that is not a spiritualist—to the possibility of extra dimensions. However, it wasn't until five years later with the work of German mathematician Theodor Kaluza that scientists began to take heed. Kaluza was taken up in the zealous attempt of his day to unify the major forces of the universe under a single theory. In a paper titled "Zum Unitätsproblem der Physik" (Regarding the Unification Problem in Physics) he lays out his discovery. If one solves Einstein's equations for general relativity using five dimensions, instead of four, then the equations of James Clerk Maxwell for electromagnetism appear spontaneously!

Jeroen van Dongen in his article "Einstein and the Kaluza-Klein Particle," introduces us to the essential history. In 1921, Einstein presented a paper by Theodor Kaluza to the Prussian Academy; it unified the gravitational and electro-magnetic fields under a single geometrical mathematics requiring five dimensions. We know that solid geometry is the study of three-dimensional shapes. So, Kaluza employed the mathematics of three-dimensional shapes to unify gravity and electromagnetism, thus concluding that electric charge is equivalent to movement in another dimension and movement in this other dimension is equivalent to charge. Thus, electric charge is understood as motion along the 5th dimension. In other words, Kaluza made charge into a metric of space!

In creating his geometrization, Kaluza stated that the components of the four accepted dimensions must not in any way depend on the fifth spatial dimension. This restriction, dubbed the "cylinder condition," was a basis for much criticism of the original Kaluza hypothesis in that he posited a fifth dimension, in order to make the mathematics work, only to negate any inherent dynamics to the dimension.

After another five years, Oscar Klein, a Swedish theoretical physicist, joined the fray using quantum analysis not available to Kaluza. He determined, in order to make the geometry work, that is, to restrict the electric charges from burgeoning to infinity, the fifth dimension in which the movement occurred had to be super small. He proposed that the fifth dimension was a circular space with a diameter of 10^{-30} centimeters. Now, our eyes can detect things only as small as .01 centimeters across. Many orders of magnitude smaller is the electron that has a diameter of about 10^{-14} centimeters, and Klein's fifth dimension is many orders of magnitude smaller than an electron. In other words, the fifth dimension is a space with no determinable extension. According to common usage of the term, and to empirical evidence, this fifth dimension is not a space at all. Rather it is what might be called a Planck scale non-space. So, for Klein, electric charge became a metric of space that is not a space; it's condition is what science now terms "compactified."

After another fifty years, scientists became involved in the unification problem on an even grander scale. Whereas Kaluza-Klein attempted to unify the theories of electromagnetism with those of general relativity, scientists of the mid-1970s began to propose the earliest versions of string theory as an attempt to unify under one theory the realm of general relativity and realm of quantum mechanics. They were attempting to unify all the known forces of the universe. And such a grand attempt at unification also demanded a grander call for extra-spatial dimensions. Kaluza-Klein required one while string theory requires at least six!

We have noted the work of Brian Greene before. He is a stellar voice among POPs lauding the potency of string theory. In his Royal Society Prize book *The Elegant Universe: Superstrings, Hidden Dimensions, and the Quest for the Ultimate Theory,* he gives a basic definition. "According to string theory, the universe is made up of tiny strings whose resonant patterns of vibration are the microscopic origin of particle masses and force charges."

Strings are thought to be in a constant state of oscillating vibration, and quantum mechanical mathematics—quantized loop spaces, infinite dimensional analysis, representation theory—can calculate the different vibrational states of strings. Oftentimes the language of music is employed by POPs to communicate with their lay audience about strings; each vibrating string plays its own note and the scale of notes composes the so-called "spectrum" of string theory. Just as Kaluza-Klein posited that movement along a fifth dimensional space appears as an electric charge in the realm of four dimensions, so, for string theory, a vibration in dimensions six through ten is detected in the realm of four dimensions as a particle, and each vibrational state appears as a different particle, each with its own mass.

For our discussion, it is important to note that two features of Kaluza-Klein theory are carried over into string theory. First, just as Kaluza-Klein was a product of mathematics, so string theory is also a mathematical creation.

Now, perhaps the most articulate spokesperson for string theory is the theoretical physicist from Stanford named Leonard Susskind. He is author of the popular book *The Cosmic Landscape,* has often been called the father of string theory and is a POPstar on YouTube, his videos often attracting hundreds of thousands of views. In an interview titled "Using Maths to Explain the Universe" he goes directly to the pith in stating that string theory is highly mathematical by nature. "I pride myself on being able to explain things to a broad audience, but sometimes you have to say it's buried in the mathematics. [The theory] just doesn't work unless you add six more dimensions to the world and we'll have to leave it at that... we're stuck using abstract mathematics."

It seems we are safe in concluding the extra dimensions of string theory are the children of mathematics, and to go one drastic scientific step farther, no POP can assuredly state that the dimensions are a factual feature of the natural world.

Before continuing, let us introduce our unique dimensional analytical tool called "Relms" to explain Susskind's assertion that science can enhance its understanding of the universe by employing a system with no empirical corroboration. Indeed, perhaps the most salient takeaway from Thomas Kuhn's *The Structure of Scientific Revolutions* is that science doesn't concern itself with truth; it concerns itself with solving problems.

Take a point mass some distance away in deep space. If we see movement in our field of view, we can either assume the point moved to the right or we moved to the left. Both describe the same phenomena and we can measure the movement, instead of with meters, with a relative meter or a Relm. A Relm is a meter as perceived from a meter away from the subject. Further, we define a Relmshell as the imaginary, yet functional spherical surface created by spanning one meter away from us. As the point moves across our view, we can measure how far it moves along the Relmshell and determine its Relm velocity as the number of Relms spanned per unit of time. (When dealing with point masses, we do not care about how far away they are from us, though the faster the Relm velocity is registered the greater the probability the point mass is closer to us.)

Let us add a second point mass and have it such that the two-point masses appear to be moving closer to each other. Using a Relmshell and a stopwatch to measure how quickly the masses are approaching one another, with the aid of kinematics and basic mathematics, we can make predictions about when and where they are expected to collide, for example. Of course, the pressing questions asked by scientists are far more complicated than this simple example, but these are the kinds of questions science is most effective in answering.

If, we are able to simulate phenomena, remaining faithful to all relevant proportions of the physical world, then we can draw the same conclusions as we did from measuring everything in that physical world. All this is to say, scientists shouldn't have an issue stripping the physical world from phenomena in order to solve the problem at hand.

So, what value does this Relms tool bring the scientist? It is minimal for this two-body problem. However if we think of a three, four or *n* body problem, by employing Relms, we can reduce the motion of *n* bodies in four dimensions (three for space, one for time) from *n* equations with ten variables (i.e. three position, three velocity, three accelerations and one time) to the motion of one body (i.e. the investigator) through an $n+1$ dimensional space. We do not propose to know what that space looks like, but it is not necessary to know in order to analyze the dynamics at hand. What these dimensional dynamics look like is irrelevant.

With Relms, scientists can no longer hold that the target of investigation is a true representation of the objective world. Investigators cannot bring themselves to believe that all objects remain static while it is they who actually whiz through n+1 dimensional space. Recall Kuhn's proposal, science solves problems; it does not speak to truth. So long as scientists keep this central to their investigation, then they'll be able to extract great understanding from Relms. Scientists confronted with the choice between solving problems or finding truth should content themselves with the fruit Relms brings: the ability to solve previously intractable problems at the expense of reality. For those interested in reviewing the mathematical correspondence between a case where bodies moving in *n* dimensional space versus one where the investigator moves through *n+1* dimensional space to simulate the movement in otherwise static bodies, please refer to the Appendix.

Now to return to Susskind, who continues by stating that there is a crucial difference between proving a system of mathematics is internally consistent and concluding that this mathematical system describes nature. He points to the various forms of geometry: Euclidean, positive curve, and spherical. Each unto itself is without contradiction, yet to say they constitute a consistent mathematical theory that describes the world is quite another thing. Without empirical support, mathematics remains isolated in its own sphere of reality.

And Susskind is supported in this perspective by most popularizers of science. Lawrence Krauss, for one, in his book *A Universe from Nothing*, opines string theory, as a vision of the universe, has been successful in dominating the minds of physicists without having once proved itself capable of solving any experimental problem about the reality of the universe. "We still have no idea if this remarkable theoretical edifice actually has anything to do with the real world."

So, string theory overall is promising, exciting, and elegantly mathematical, but without empirical help, it is yet many orders of magnitude from the unified theory of the universe.

The second feature carried over from Kaluza-Klein to string theory is that extra dimensions, as demanded by the mathematics, are purported to be super small. In the case of dimensions six through ten, they are purported to be many billions of times smaller than that which has been observed in particle collision experiments. And this nano-small, mathematics-based reality of string theory's extra dimensions presents a particular challenge to the POPS. Recall that they, in attempt to speak to intellectual lay readers, eschew mathematics and rely on argument, images, and stories to convey their version of the scientific message. So, how do they speak and write and debate the extra dimensions? They turn to an analysis of the second demand witnessed in the POPs canon; they seek extra dimensions in the very small because they are of the view that human beings do not perceive them.

Almost all POPs reference the un-observability of extra dimensions. Brian Greene, in *The Elegant Universe*, holds "String theory also requires extra space dimensions that must be curled up to a very small size to be consistent with our never having seen them." M. Seifert, in a lecture titled "Calabi-Yau Compactification" warns, "If one wants to be able to apply the elegance of string theory to explain the observed world, however, one has to come to grips with the fact that the world we observe is most emphatically not ten dimensional..." Lawrence Krauss, in *A Universe From Nothing*, believes "After all, the world of our experience is not ten-dimensional, but rather four-dimensional."

So, the accepted fact among POPs that extra dimensions are not perceivable has led them to seek these dimensions as simply a miniaturization of the three dimensions of their daily experience. Now, before critiquing this un-observability demand from the perspective of Nihilology, we would like to reference the reflections of two POPs who have tentatively posited extra dimensions that are not nano-small.

In the late 1990s, Lisa Randall, the Harvard theoretical physicist who we have discussed several times earlier, had an outsight, as witnessed in her book *Knocking on Heaven's Door*, on how to approach the problem of extra dimensions in "an orthogonal way." "Orthogonality" means "at right angles" and is an apt choice here in that Randall began to seek an extra dimension along a vector that was not a miniaturization of the daily three but at a right angle to the common vector. The extra dimension she found is where gravity roams, and she holds it can exist on a large scale.

Now Randall has translated the so-called "warped geometry" of this extra dimension into clear language for her lay audience. It is accepted physics that the other three forces of the universe are carried by their unique particle: electromagnetism by photons, strong force by gluons, weak force by bosons. Thus, it is surmised that gravity is also carried by a unique particle, deemed the graviton. Randall tells us the graviton is the expression of a string vibration as it appears in the realm of our daily three spatial dimensions. The utter weakness, may we say the quiet whisper, of this gravity string vibration in the extra dimension has its counterpart in rarity of expression as a graviton in the daily dimensions, so rare as to go unnoticed. In her article "Theories of the Brane" Randall says, "The physics far away is in fact so entirely irrelevant that the extra dimension can be infinite, with absolutely no problem from a three-dimensional vantage point."

There is one problem, however, that remains in this proposed extra dimension. Since scientists need to look, with their eyes or via magnifying mechanisms, using photons and photons are larger than gravitons, they could not see them in any case, even if it could be determined that a visitation of a graviton from the extra dimension had occurred. Thus, though Randall frees us from the stricture of compacted nano-small extra dimensions, we cannot detect the presence of the string for gravity in our three-dimensional space and again thus cannot confirm the extra dimension. So, the existence of this particular extra dimension, this realm of gravity, is still not an empirical reality.

Now, whereas Randall exhibits a rigorous, orthodox method of physics to seek out an uncompacted extra spatial dimension, David Bohm sought the same via a methodology of universalism with a flamboyant philosophicality. He prophesied that systems cannot be analyzed as an interaction of independent particles with the advent of quantum theory. Rather particles must be understood as projections of a higher dimension of reality.

Bohm, in his book *Wholeness and the Implicate Order*, makes use of an image, of a type we have become familiar with in the POPs canon, that attempts to illustrate how higher dimensions would be experienced from the viewpoint of lower dimensions. Bohm has us imagine a cubical fish tank with two separate cameras, each focused through adjacent walls of the tank on the fish swimming about inside. The scenes shot by the cameras are viewed on two screens in a separate room. Observers will see different swimming moves on the part of the fish, but the movements will always be correlated with the other.

Of course, we, being aware of the design of the experiment, know that the two images do not refer to independently existing actualities. "This actuality is of higher dimensionality than are the separate images on the screens; or, to put it differently, the images on the screens are two-dimensional projections (or facets) of a three-dimensional reality."

Ultimately, Bohm proposes that seemingly independent phenomena in the manifest universe are expressions of a higher unseen, perhaps unknowable, dimension wherein the independent phenomena are unified. He calls this dimension the "Implicate Order," and it is of infinite size.

These two POPs are outliers in proposing an extra spatial dimension that is not nano-small. They do a great service that should exert an effective nudge upon the vector of string theory research. However, these two outliers still cling to the ground notion within the POPs canon that extra dimensions are not observable, even though the reason is not their nano-size, and that the extra dimensions lie outside our world of perception.

Now it is time to bring forth the promised perspective of Nihilology which is in accord with Randall and Bohm in holding that the extra dimensions need not be nano-small. But beyond that, we propose a perspective on extra dimensions that is both more radical and more mundane. For, according to Nihilology, extra dimensions are observable, and they reside within the manifest physical world!

Note these two POPs statements that will help us focus the introduction to our argument. The first is from Brian Greene in a section titled "The Illusion of the Familiar" from his book *The Elegant Universe*, "Experience informs intuition. But it does more than that: Experience sets the frame within which we analyze and interpret what we perceive... [it] determines our interpretive mindset."

The second quote is a rather famous one from Robert Oppenheimer. "There are children playing in the streets who could solve some of my top problems in physics, because they have modes of sensory perception that I lost long ago."

Greene warns us that how we perceive the world determines the vector of our scientific research and frames how we interpret the data found in that research. Oppenheimer stands with this perspective but adds a note of nostalgic optimism. Nihilology agrees both of these physicists, for we will now present an argument that our perception of the physical world determines our research findings and, with a bit of childish optimism, we can perceive more dimensions of the physical world with a facility that arises from a deeper understanding of the neurology of the eye as a tool employed by a variably located self.

Now, it is widely accepted by neurological science that, of the five senses, vision is by far the most important in our human interaction with the physical world. It is seeing that reveals dimensionality and seeing is much more a creative selective act than a camera on the world. For example, we are natural-born experts in projective geometry who can conjure a three-dimensional world out of two-dimensional snapshots that fall on our retina. And, as we discussed earlier regarding the work of Julian Jaynes, all this conjuring takes place without us being at all conscious of what is occurring within us.

Many years after garnering the Nobel prize for his work with James Watson in discovering the double helix of DNA, Francis Crick wrote a book titled *Astonishing Hypothesis: The Scientific Search for the Soul.* He recaps the science of human seeing by calling it an active, constructive process. Humans require a multi-valent reading of a visual scene in order to see it. "It is difficult for many people to accept that what we see is a symbolic interpretation of the world..."

We should note here that Crick's methodology of seeking out the nature of the soul via an analysis of vision played a role in the birth of Nihilology's method for this theme. We were taken by his basic premise that there is a correlation between human vision and human "soul."

We then evolved this premise to use seeing as a means to analyze the human self. The soul remained, even for Crick, a just-beyond-knowing entity of a faithed nature whereas the self, in our analysis, is a variably located entity, empirically verified by research in neurology and neurological psychology.

So, while Crick brings out the creative power of seeing, Frank Wilczek highlights its selective powers by referencing the shadow watchers in Plato's cave who only saw a smidgen of reality. When we look out at the world, we miss much more than we catch and need complex inferences to glean the visual information we think we need. "... the information we sample about the world is both very partial and very noisy... [we see] only paltry samples from the cornucopia of information that the world puts on offer."

Crick and Wilczek together provide us with fertile hints as to the creative selective nature of human seeing, but it is in the work of Donald Hoffman, Quantitative Psychologist at UC Irvine, that this science comes to light clearly. Hoffman is a POPstar whose 1998 book *Visual Intelligence: How We Create What We See* was read widely among lay intellectuals. His TED Talk video titled "Do We See Reality?" cited here has received over 500,000 views on Youtube. Again, we note the cultural impact of these popularizers of science. (Also, we might note how appropriate it is for the deepest impact on the popular science of human vision be a format that is visual!)

Hoffman does a dance on the TED stage. He uses high impact images in step with somewhat folksy language to effective ends in convincing an audience of lay intellectuals. He begins by lamenting aloud how badly we misinterpret our experiences. "It feels like we are taking a snapshot of the world around us but the fact is... Neuroscientists tell us that we re-construct the reality around us..."

Notice that neuroscience has established that seeing is a re-construction of the world. Also notice the "re" here; Hoffman maintains that the physical world is a truly existent objective reality, objective in the sense that it exists outside the mind and body of the human being. Neither Hoffman nor Nihilology proposes any form of idealism.

Hoffman continues, "Once we let go of our massively intuitive, but massively false assumption about the nature of reality... dare to recognize that perception is not about seeing truth, it is about having kids." It is a "massively false assumption" that human vision is about seeing the truth of the objective world by taking a trustworthy photograph and projecting it on our retina. Rather, seeing has served an evolutionary mission, a life hack as Hoffman calls it, created to aid in keeping us alive and "having kids."

And finally, Hoffman creates a classic image in the realm of scientific metaphors. He suggests that the objective world of human vision **is**—not **is like** as with a simile—a 3D desktop inhabited by icons. This notion, a neurological fact, is essential to our argument regarding research into the extra dimensions required by string theory. Indeed, the objective world is an objective **space**, a stage upon which objects of various qualities appear to human seeing. And, as we will clearly show, depending on where we look from, that is, where the variously located self resides, different essential qualities of objects will be perceived. "[With seeing] evolution has given us an interface that hides reality and guides adaptive behavior... a set of perceptual symbols that are designed to keep us alive... reality is more like a 3D desktop... Space as you perceive it is the desktop and physical objects are the icons in that desktop..."

Recall our discussion before of Lisa Randall who, in seeking the infinitely large extra dimension of gravity, worked in a manner that was orthogonal, at right angles to the vector of common research. She stated that she needed to think outside the box of conventional POPs research. We would like to suggest that Randall also needs to think **inside** the box in that the objective world is a space or stage of more dimensions than she assumes. Seeing with the eyes, done right, reveals extra physical dimensions. And we might add, along with Hoffman, that seeing with the eyes done right may serve an evolutionary function as humans are forced to stay alive in a unified quantum/relativity world.

So, this respected panel of popularizers of science—Greene, Seifert, Krauss, Randall, Bohm, Oppenheimer, Crick, Wilczek, Hoffman—have taught us much about the extra spatial dimensions required by string theory and the science of human seeing. We have learned, while the POPs canon is dominated by the proposition that extra spatial dimensions will be discovered at the super small end of the continuum of the material world, certain researchers are seeking larger extra dimensions. We also learned that human seeing is a creative selective process, that "we re-construct the reality around us," and we glean "paltry samples from the cornucopia of information that the world puts on offer." Each of these POPs hold to their individual perspective on these areas of research, yet there is one perspective they all agree on: human beings do not wield the power of seeing extra dimensions. As Brian Greene declares, "Yet there are certain things that we *all* experience... [it is an] apparent fact that we *see* precisely three spatial dimensions."

Now we ask our readers to recall our earlier analysis wherein we presented a topology of knowing. The science of epistemology posits three sorts of truth: objective, subjective, and intersubjective. We asked the question: where do these forms of knowing reside? Objective knowing resides in the world outside the individual and is verified empirically. Subjective knowing resides inside the individual and is verified internally via epistemic feelings. And intersubjective knowing resides in the social realm and is verified in relationships. As we stated regarding intersubjectivity, "The views of the world of all the members of a group form into a common sense, usually unreflective, complex of meanings that make up the basis of a shared language and a shared experiential context that Husserl called a "homeworld" [*heimwelt*]. The homeworld is a consensus reality held together by interlocking conventions that make the people in the group aware of what is normal and serves as a source of epistemic confirmation for everyone."

We propose the POPs conviction that humans are constrained to seeing a world of three spatial dimensions is a form of intersubjective knowing, not objective knowing, as they univocally aver. This critique leads directly and conclusively from our analysis of neurological science that reveals seeing to be a creative selective process. The experience of seeing is a conjuring performed by the self or mind as we or it selects relevant data out of a vast spectrum that enters the eye from the world. Thus, the realm of verification for vision is not the objective world as it might be were our eyes taking a photograph of the world and displaying it on the retina. It is not the subjective realm of individual epistemic feelings as it might be if we lived in an idealist's world. Rather the realm of verification for seeing is relationships among a thought community; it is an agreement among POPs as to the nature of their homeworld!

It is our contention that the accepted intersubjective agreement that human beings perceive only three spatial dimensions, either via our senses or with the aid of technology, in contrast to the demand that string theory requires six extra dimensions leads popularizers of science to seek the unseen dimensions within the material world, but at nano-small scales in the realm of quantum mechanics. POPs accept their daily level, the Newtonian realm we might say, of visual spatial experience. They hold fast to their intersubjective notion that three-dimensional visual experience is universal, that is, objective truth.

Furthermore, POPs also reject any sort of world that is not material, as we have discussed earlier, and as witnessed in these disparaging remarks about mysticism by Brian Greene, "The suggestion that our universe might have more than three spatial dimensions may well sound fatuous, bizarre, or mystical."

In the end, popularizers of science are stuck with seeking out the extra dimensions demanded by string theory on the extreme small end of the continuum of materiality. This research vector may be successful at some future time in empirically locating extra spatial dimensions, but this research strategy is motivated by a fundamental fallacy. It is the fallacy of POPs' intersubjectivity, and it delimits their experimental designs. Extra spatial dimensions, in fact, can be found in our very realm of daily visual experience. Nihilology will now put forth an argument for how this is so. It is fairly complex and a fairly long path, so please walk along with us, with a bit of patience.

Let us begin by delving into a few basics of what neurology tells us about the relationship between the brain and the rest of the body. For instance, Rollin McCraty, psychophysiologist and Director of Research at the HeartMath Institute, in an article titled "Heart-Brain Neurodynamics," points out the brain is hard-wired with bi-directional neural connections between the cognitive center (cerebral cortex) and the emotional center (amygdala), with the majority of those neural connections carrying information **from** the emotional center **to** the cognitive center. For our discussion it is important to realize that this directional dominance also holds true for the neural highway between the heart and the brain which is constituted of as many afferent fibers as there are efferent connections. "Remarkably, we now know that the heart sends more neural traffic to the brain than the brain sends to the heart."

McCraty's findings are consistent with the research of many others in the field of psychoneurology, and for us it has one intriguing implication. The heart is not mainly a reflector of cerebral, or even brain-centered emotional activity, but is the originator of emotional perception and experience.

Now the predominance of afferent neural traffic is even more robust in the case of the stomach. It was quite a jolt for scientists, raised with a bias of brain-over-body dominion, to discover that more than ninety percent of the traffic on the vargus neural highway between the gut and the brain was headed to the brain while only ten percent traveled to the gut.

Recent studies have shown, whereas the brain is quite dependent on neural information from the gut, the gut can self-direct almost all of its functions without any reliance on the brain whatsoever. Of late, this complex system of neurons has been legitimized in popular culture as a quasi-independent center, witnessed by the avid readership of Dr. Michael Gershon's 1998 book titled *The Second Brain.*

These research findings help us take a few important steps in our discussion because neural centers of the body outside the head, especially the heart and the stomach, play essential roles, indeed underlie, our experience of selfhood.

Let us now turn to the ground-laying research of Wilder Penfield, the American-Canadian neurosurgeon who worked for decades to create a neural map of the mind and body. He became a POPstar as a perk of his homunculus, a grotesque humanoid character with bulging hands and lips that he crafted based on the mass of brain tissue involved in sensing various bodily functions. And, though homunculus was a sort of light-hearted meme of his day, Penfield was serious in his choice of name for his meme. The Latin term "Homunculus," originating in European alchemy of the 1500s, translates as "little man" and refers to a miniature person inside the gross human body. And the search for an internal "little person" was a life-long zealous fascination in Penfield's medical practice and writings.

Throughout his scientific career Penfield, like many other scientists, strove to prove that the brain is the sole basis of what humans experience as the mind. But when there was a buildup of research available, he wrote a book-length answer, titled *The Mystery of the Mind*, to his essential question, "Can the mind be explained by what is now known about the brain? If not, which is more reasonable of the two possible hypotheses: that man's being is based on one element, or on two?"

Beginning in the mid-twentieth century, Penfield pursued his question as to the single or double nature of the human being via a research methodology on live human subjects that would make university institutional review boards greatly distraught today. He stimulated with electrodes various brain regions of awake human subjects, all the while speaking with the subject about their experience. Over years of applying this method, Penfield realized a subject could experience an electrode-stimulated memory, taste, or vision and verbally reflect on that experience at the same time. This led him to conclude that humans have a place deep within or high above the brain where another awareness lies.

Pursuing this question even farther, and this was the technique of awakening for Penfield, he began to stimulate areas of the brain that caused a subject to make physical responses, such as clenching their fist. At the same time he would request the subject to try to relax their fist; then their fist didn't clench so tightly! For Penfield this was clear demonstration that the person whose fist was clenching and the person who was working to ease the clench were not one and the same! The person who was easing their own bodily clench, the person who he was talking to, Penfield called "The Watcher."

The basis of a watching center outside the brain is lent credence in a wide range of scientific neural coding research. A popular journal article titled "Neurology and the Soul" provides us with an approachable summary of research regarding the coding of visual perception. It states that all visual information is encoded in complex sequences of electrical responses in the highest level of visual cortex. But, therein lies the catch: the brain is only capable of encoding visual information. "There must be an 'I' distinct and aloof from the physical brain that interprets the code. When we look at an object, we perceive the object. We don't perceive sequences of electrical charges ... There is, therefore, a non-brain entity that translates patterns of electrical changes into conscious perception."

The notion of personal self as watcher aids us in taking yet another step toward a full elucidation of the human capacity to perceive extra physical dimensions. Indeed, we employ the notion quite literally in our argument below.

The notion abounds in the literature of religion and philosophy as well. Listen here to Gary Vey, an investigative researcher and creator and editor of *Viewzone Magazine*, wherein he pursues a popular science agenda according to a methodology akin to the pattern of knowledge of Nihilology. "Self-hood happens to us, delayed and filtered. In truth, the self is the watcher... 'Self' is a concept used to designate the entity which witnesses our life."

To go another step, the selfhood of human beings is much more than a concept. It is a visceral experience in both the connotative and neurological sense, held deeply as an epistemic feeling and located in the "viscera" or organ nerve centers. The research into the self sense is complex and widespread among sub-disciplines of neurology and psychology. There is, however, clear consensus on the basics. As Jakub Limanowski and colleagues conclude in their article titled "Where Do We Stand on Locating the Self?", "Moreover, results reveal that people believe there is one single point inside the human body where their Self is located."

Now, whereas humans can quite easily and clearly locate their self in their body, that location varies between individuals and within a single person according to their differing internal states. The well-known science writer Emma Marris focuses the diffuse research on self location in an article titled "Where is Your Self." She opens the article with an amusing quip, "First of all, my self is definitely in my head. I've got a little homunculus sitting behind my eyes, running the whole show, like Captain Kirk on the bridge of the Enterprise."

This statement is revealing of three important features of the research on self location. First, most people in the West locate the self in their head. Second, Penfield's meme of the homunculus is alive in the lived experience of human beings. Finally, and this is central to our discussion of the observability of extra spatial dimensions, Marris senses her self as located "behind my eyes." As we will show, the self is located behind the eyes even when the self is experienced in the body below the head.

Marris hones in on a set of research findings from 2008 led by a group of Italian psycho-neurologists who define the self as the "I-that-perceives." She then concludes thus, "So it seems the 'I' spot can change depending on the context. That makes sense... It could be where our thinking is going on, or where our emotional center is. Alternatively, it could be the point from which we initiate action..."

Before continuing, let us point out that the modern research on the variously located self reverberates a classic Greek teaching on the tripartite nature of the soul. For example, Plato distinguished three qualities of the soul: reason, passion, and appetitive drives. These three qualities he seated in the head, the chest, and the belly.

Now these three locations of self, as documented by Marris, are by far the most often reported by subjects in psychological research studies. And the standard deviation of reports remains constant throughout research designs ranging from self-report to projected body images to blindfolded pointing. However, it is not to be concluded there are physiological differences that account for why some people locate the self primarily in the head, others in the heart, and still others in the stomach. The main variables that determine self location, as reported by Marris above, are the contexts of internal states occurring for a person at the time of report.

Readers should recall these locales of self correlate significantly with the three quasi-independent neural centers discussed before: brain, heart, and stomach. This correlation lends great credence to the argument that these neural centers provide a physiological bases for the experienced locations of the self.

Just as an aside here, there are also cultural differences that affect where people locate their self. The Japanese, for example, choose the stomach (*hara*) much more often than Europeans. The phrase *hara no aru hito* literally translates as a "person with a stomach," but has the connotative meaning of "one with a center."

In sum then, research shows that the self is variably located. It is most commonly experienced in the head, the heart or the stomach. These give location to three central human experiential aspects of selfhood: concept, feeling, and will. And, as we shall show, these correlate with three specific sorts of experience of the objective world: associations, aesthetics, and dynamics.

Now, in order for readers to appreciate the import of the variously located self in determining our visual experience of the objective world, we will return to the task of model making. We earlier created a model of outsight as a way of explaining the POPs oft-cited, but never analyzed, eureka! experiences leading to new theories and research pathways. We then dovetailed in a model of awe as a way of explaining the POPs oft-cited source of inspiration for their entering into the field of physics and source of vitalizing amazement at the cosmos. Now we build yet another model to dovetail in order to clarify the argument of Nihilology that certain extra spatial dimensions are material and observable. We call it Four-Staged Kenosis.

In at least three earlier discussions we have employed our heuristic that we tagged as the rarefying principle. The principle has shown itself in the history of science thus: the function of created things is more and more revealed by the shedding of excess matter. As a created thing becomes more and more understood, it turns out to be less and less dense; it sheds matter, becomes more rare.

Our rarefying principle was also witnessed in the process of *kenosis* [Greek "emptying"] wherein the self is stripped of its properties unto Nothing. In the earliest theological usage, *kenosis* referred to the need of Christ to empty himself of his divine nature in order to incarnate as a human being in the form of Jesus. The term evolved to mean the emptying of individual selfhood toward the greeting of the Nothing of the person with the Nothing of the divine. (Recall that the *via negativa* can also be understood as an expression of the rarefying principle, i.e., stripping all conceptions from the actuality of the divine.)

We discussed historical perspectives on the *kenosis* process. The Beatitudes of the Gospels give praise to the non-inflated person, one who empties themselves and, in doing so, allows the Holy Spirit to enter. The poesy of Meister Eckhart intones, "If I wish to write on a white tablet… I must wipe the tablet clean of everything, and the tablet is most suitable for my purpose when it is blank." And in the writings of Carl Jung, emptying intends an internal psychological process of cleansing the ego of tendencies that blind a person from their deeper relationship to the Self.

Now, Nihilology will respectfully borrow this term *kenosis*, importing many of its connotations, to give title to our new model: Four-Staged Kenosis. This model will serve as a heuristic, a tool for understanding exactly how the variously locating self gives place—head, heart, stomach—to a process of rarefying selfhood. These locales in turn correlate with three internal experiences: thinking, emotion and will (with a fourth being beyond internal experience). They also correlate with three objective world perceptions: associations, aesthetics, and dynamics (with a fourth being beyond dimensions.)

Our readers might recall that popularizers of science hold tight to the basic belief that the universe must be explained as a natural phenomenon. No divine agents can be called upon when mysteries crop up. In keeping with this POPs belief, we must note that, while importing certain evocative notions from the theological term *kenosis*, our Four-Staged Kenosis model, as it applies to extra dimensions, does not describe a religious or spiritual process in any sense of those words. While the process described in our model is by nature dependent on the reality of Nothing, i.e., the realm of the divine, the model is rigorous in calling upon POPs to use their physical eyes from the location of their experienced self to perceive objective spatial dimensions beyond their intersubjective three. This will aid us in taking a few steps toward answering the demand of string theory.

Before going farther on our path to understanding observable extra dimensions, let us fill out our dovetailed model of outsight, awe, and Four-Staged Kenosis. Earlier on we concluded that the common four-staged model of insight is superficial in that it fails to recognize the role of trans-heuristic images and it is silent regarding the workings of second stage of the model, coined as conceptual rest, relegating it to the realm of the "un." Thus, we suggested that scientific insight is more subtly and potently described as a 6-staged process of outsight.

Six Stages of Outsight

1: Intensive Thought
- employment of images

2: Waning Thought
- left brain » right brain

3: Meta-concept
- parallel processing

4. Nihilitive Knowing
- inperience

5: A-ha! Outsight

6. Verification

We then offered an analysis of scientific awe, reported throughout the POPs canon as occurring while gazing at the night sky and giving "secular" inspiration to their careers in cosmology and astrophysics. The past decade has witnessed a fascination with awe among psycho-neurologists, and they have come to consensus that awe is awakened when three aspects are coincidentally active. First, perceptual vastness is presented, as is the case when a scientist is working intensively on a problem in the evening while under a clear dome of the firmament. Second, extra-conceptual neural activity is dominant as expressed in a shift from left brain rationality and analytical language to right brain emotion and holism. Third, there is a waning sense of self.

Our analysis of awe dovetails nicely with our six-staged model of outsight. Both awe and outsight are liminal experiences regarding the realm of Nothing. They arise on the threshold of the infra-conceptual and the super-psyche. Awe arises on the ingress. Awe is what it feels like to approach the realm of Nothing. Outsight is what it thinks like to re-emerge from the realm of Nothing. We could say awe is the feeling aspect of liminality. Outsight is the thinking aspect of liminality.

As we said before, Nihilology is quite in accord with popularizers of science in averring that awe and outsight are not *per se* spiritual experiences: awe and insight both arise on the threshold of the realm of Nothing, the realm of the divine. Where Nihilology advances over the POPs is to acknowledge Nothing. Nothing as a declarative with a pull to it, affirmed beyond the threshold where awe and outsight find their élan. Nihilology also holds that awe and outsight are fully dependent on this liminal relationship with the realm of Nothing. Awe and outsight would not arise were it not for Nothing. Nonetheless, recall that popularizers of science are not prone to the disciplines of staying awake, thus their entrance into the realm of Nothing, their passing through, will be absent awakeness, carrying with it no inperience at all.

Six Stages of Outsight/Awe

1: Intensive Thought

- employment of images

2: Waning Thought/**Awe**

- left brain » right brain

3: Meta-concept

- parallel processing

4. Nihilitive Knowing

- inperience

5: A-ha! **Outsight**

6. Verification

Now to dovetail in our Four-staged Kenosis model. Psychoneurological research shows that human beings most commonly experience the self in the head, the heart, or the stomach (which perfectly embody in the three quasi-independent neural centers). These give location to three central human experiential aspects of selfhood: concept, feeling, and will. We proposed that our outsight model gives witness to the rarefying principle regarding human conceptual activity: rarefying from intensive rational thought to ebbing of thought accompanied by a shift from left brain to right brain activity. Then rarefying down into the infra-conceptual wherein unconscious parallel processing occurs. Rarefying of thought sublimates in the realm of Nothing wherein thought has no life and Nihilitive Knowing arises.

It is the unique proposal of Nihilology that our Four-Staged Kenosis model describes the very same rarefying principle, but regarding the self. We hold that the outsight model and the kenosis model are utterly parallel to and amplifying of each other. We further hold that, just as thought rarefies into Nothing of Nihilitive Knowing just so self rarefies into emptying (*kenosis*). Both tend toward inperience of Nothing, one below the continuum of concept, the other below the continuum of identity.

Six Stages of Outsight/Awe	**Four Stages of Kenosis**
1: Intensive Thought • employment of images	Self located in head • thinking
2: Waning Thought/**Awe** • left brain » right brain • ebbing self	Self located in heart • emotion
3: Meta-concept • parallel processing	Self located in stomach • will
4. Nihilitive Knowing • inperience	Kenosis of Self
5: A-ha! **Outsight**	
6. Verification	

Thus, we present our dovetailed models of outsight, awe and Four-staged Kenosis in graphic form. But, readers may wonder, where is the justification that the variously located self gives witness to a process of kenosis of self? This proposition is attested to throughout our earlier discussions regarding Nihilitive Knowing and the disciplines for staying awake in the realm of Nothing; it is quite straight forward to come to this conclusion by a simple screening of these arguments already put forward, looking at them with the new eye of self emptying.

Earlier we argued that there is a precise confluence of notions from across the religious traditions of the East and West. We analyzed nine traditions, all giving voice to a single notion: the human being must empty themselves of their self in order to unify with the most rare divine as Nothing. Here are four examples from among these to view with this new eye:

Recall, in Taoism the classic discipline for self emptying is called *zuowang* or "sitting-forgetting." Designating the discipline thusly carries a bivalent meaning. It points to the ebbing of concepts as self begins to rarefy as well as the overall discipline of emptying selfhood. Look again at this passage where a student named Yan Hui is talking with his master Zhongni:

> Yan Hui said, "I am making progress." Zhongni replied
> "What do you mean?"
> "I have ceased to think of benevolence and righteousness," was the reply.
> "Very well; but that is not enough."
>
> Another day, Hui Hui again saw Zhongni, and said, "I am making progress."
> "What do you mean?"
> "I have lost all thought of ceremonies and music."

In that we are looking closely with our reinspection, let us note it is more accurate to translate with the connotative "stopped focusing on" rather than the literal "ceased to think" or "lost all thought." With this in mind, we note that "benevolence and righteousness," according to the urtext *Analects of Confucius,* refers to governance via compassion and shame, i.e., effective emotions. Also, ceremonies were an essential aspect of structuring the society and represent the embodied acts of human will. Thus, with Yan Hui's discipline of sitting-forgetting, he rarefies from thinking to effective emotions, and then further rarefies into willed acts. This rarefication fits nicely with stages one through three of our Four-taged Kenosis.

Please note Yan Hui continues to sink out the bottom of identity altogether toward the inperience of Nothing (*wu*) which, as we are arguing, correlates with our fourth stage of kenosis. We will not focus on this fourth stage overly much in that it is below the continuum of objective experience. And objective experience of extra spatial dimensions is our challenge for explication in this final theme of this section.

We also analyzed the rarefying of the self according to teachings of Hinduism's non-dual school as expressed in *The Crest Jewel of Discrimination.* It holds that the dance of illusion (*maya*) misleads us to believe that the *koshas* constitute our real identity when in fact they are sheaths that cover over the actual nature of selfhood. With the stripping away of these illusions from self, reality is greeted. In our earlier discussion we quoted verses on the sheath of the physical body "made up of fat, flesh, feet, and lesser limbs." The verses of *Crest Jewel* continue by rhapsodizing about three others made up of rational activity, sensual experience and vital breath:

The mind-formed vesture is formed of the powers of perception and the mind
it is active in making a distinction of names and numbers
Therefore mind is a cause of man's bondage
mind ceaselessly shapes the differences of body, of color, of condition, of race

They who, fooled in these sensuous things
are bound by the wide noose of lust, hard to break asunder
Sensuous things are keener to injure than the black snake's venom

These also are the life-breaths
Up-breathing, down-breathing
the forward moving of breath, and the outward moving
these are the doings of the life-breaths

Rationality, in *Crest Jewel* poesy, is portrayed as the "mind-formed" sheath that analyzes the world into names and numbers. Emotions of the heart are talked of in terms of lusting after the sensate objects of the world. The acts of will are elaborated—part of a complex and sophisticated first-person science—as bodily movements whose force originates in the vital breaths. With the discipline of staying awake, these illusory sheaths are progressively emptied of identity. For our current discussion it is essential to note that these three emptyings exactly jive with the first three stages of our Four-staged Kenosis model.

We then turned to the traditions of the West, first considering two voices from Christianity who intoned the rarefying of self as a condition for greeting God as Nothing. St. John of the Cross said in *The Dark Night of the Soul*:

This darkness should continue for as long as is needful
in order to expel and annihilate the habit
which the soul has long since formed in its manner of understanding
[the dark night] impoverishes it and empties it
of all natural affection and attachment

These verses give reference to emptying the supplicant of cognitive functions and emotional functions. Again, these verses clearly echo stages one and two of our Four-Staged Kenosis model.

And Meister Eckhart preaches a clear doctrine for emptying in accord with the second stage of our model:

Similarly God works in all hearts not alike
but in proportion as He finds them prepared and susceptible
If the heart is to be ready for the highest
it must he vacant of all other things.

Regarding Judaism's discipline toward self emptying and unifying with the divine Nothing (*ayin*) we discussed the contemplation of the *sefirot*, presented as a trans-heuristic image of a ten-limbed tree of life. The *sefirot* are the ten emanations of the divine *ayin* (God) manifesting in the world. Human beings contemplate the *sefirot* as an ascending hierarchy toward the understanding of God. The actual greeting of God as Nothing occurs beyond the continuum of the emanations and this greeting can only take place through the forgetting of self, a stance akin to the sitting-forgetting of Taoism mentioned before. As Issachar Ber of Zlotchov averred, "... through realizing that there is no place empty of [Yahweh]... " Again we see this bivalent meaning of "forgetting," intending both the rarefying of concepts as well as the overall discipline of emptying of selfhood.

And what is important for us to note in our discussion of the rarefying of self, the *sefirot* are also presented in a trans-heuristic image of a ten-aspected human body. The contemplation of the *sefirot* is, at one and the same time, an ascent of characteristics of God's manifestation and a descent of the human body, as intoned in George Robinson's "What are the Sefirot?"

Keter is the area just above the head, what might be called the aura of the body
Hokhmah and *binah* are the hemispheres of the brain, the region of thinking
Hesed and *gevurah* are the right and left arms of judgement and mercy
Tifiret is the upper torso or heart region of compassion and pity
Netzah & *Hod* are the left and right leg
Yesod is the penis of fertility
Malkhut is the feet which ground a person on the earth

The Kabbalistic discipline for staying awake is to contemplate the *sefirot* of the body toward a forgetting of self. This contemplation brings about a forgetting of thinking in the form of emptying of *hokhmah* and *binah* as well as a forgetting of compassion and mercy in the form of emptying of *tifiret*. These give witness to stages one and two of our Four-Staged Kenosis model.

Bolstered by this review of the traditions of East and West, we are confident in our unique analysis that the variously located self, as presented by the research of psychoneurology, gives witness to a process of kenosis of self. The experienced self does not migrate haphazardly but is rarefying as it descends from head to heart to stomach, accompanied by a kenosis of concepts, emotion, and will.

The kenosis process, as it applies to the experience of extra spatial dimensions, is not of itself a religious or spiritual experience. We are merely guided by our methodology of employing the pattern of knowledge from the writings of religion, philosophy, and science. Herein we call on the teachings of religion as they are, in this area of knowledge regarding emptying of self, most sophisticated in its explication.

Remember, it is psychoneurology that posits the variously located self and, in the case of awe, the human experience of waning selfhood.

Also recall that for Nihilology there is no religious experience *per se* in that the divine is met only when the self of the human has rarefied to Nothing and the divine has rarefied to Nothing. This greeting has no distance. The divine is not objective to the person, and their greeting is more properly referred to as inperience.

Now, having walked this long path of ground-laying, we are prepared to answer our central question of this theme: how is it possible to discover extra spatial dimensions, as demanded by string theory, in our very realm of daily visual perception of the objective world?

Recall that psychoneurology points to a non-brain entity that translates patterns of electrical changes into conscious perception. This research gives a neural basis for the widely accepted view among scientists that the act of seeing is a creative selective process. Recall also the concomitant conclusion of neurosurgeon Wilder Penfield that humans have a place ancillary to the brain from where we witness our perceptions. This self as watcher is also key to answering our question of extra dimensions.

Indeed, it is the contention of Nihilology that the eye must be treated just like a microscope or particle sensor, that is, as a tool employed by the self to do its extension into the objective world. Further, the self uses the tool of the eye differently, and sees the world differently, according to its location in the head, the heart, or the stomach.

This notion of **using** the eyes is widespread in the canons of psychology and philosophy. For instance. Peter Michaelson, an oft-read depth psychologist, in his article titled "When Eyes Are Blinders of the Soul," advises that people are not able to perceive their world wisely unless they develop a reflective life "which sees from the depths *behind* our eyes. When we see in such a manner—from behind our eyes—we're more conscious of how to resist being lured visually into the arms of negative emotions."

Hear also William Blake in his epic poem "Songs of Innocence:"

This life's dim windows of the soul
Distorts the heavens from pole to pole
And leads you to believe a lie
When you see with, not through, the eye

This seeing through the eyes is also not an unknown stance in the canon of science. Recall Robert Oppenheimer's nostalgic quote about the vision of children, "There are children playing in the street who could solve some of my top problems in physics, because they have modes of sensory perception that I lost long ago." Oppenheimer, as an aging adult theoretical physicist still had the same eyes that he had as a child. And his vision had not degenerated in the optometric sense; it was his skill in using the eyes that had degenerated.

So, when we use the eye as a visual-perception research tool for the self variously located, what do we see in the world? It is the perspective of Nihilology that the self located in the head experiences physical objects imbued with associations in the form of names, memories, and theories. Self located in the chest experiences physical objects imbued with aesthetics in the form of light, shape, and color. And self located in the stomach experiences physical objects imbued with dynamics in the form of forces, momentum, and movement.

Six Stages of Outsight/Awe	Four Stages of Kenosis	Imbued Seeing of Objects
1: Intensive Thought • employment of images	Self located in head • thinking	Associations
2: Waning Thought/**Awe** • left brain » right brain • ebbing self	Self located in heart • emotion	Aesthetics
3: Meta-concept • parallel processing	Self located in stomach • will	Dynamics
4. Nihilitive Knowing • inperience	Kenosis of Self	"Below" Objects
5: A-ha! **Outsight**		
6. Verification		

Thus, we present in graphic form our dovetailed models of outsight, awe, Four-staged Kenosis and associated visual experience of objects. But again, our readers may wonder, where is the justification that the variously located self, coursing down through process of kenosis, sees the objects of the world in a correlated creative-selective manner?

Let us start at the top of the body and at the beginning of time. The Hebrew Bible, Genesis 2: 19-20 says:

> So out of the ground the Lord God formed every beast of the field and every bird of the air

and brought them to the man [Adam] to see what he would call them and whatever the man called every living creature, that was its name

The man gave names to all cattle, and to the birds of the air, and to every beast of the field

Note that Adam had in-born full capacity with thought and language. He did not need to be schooled by God. Also, he had memory in that he had to name a host of birds and animals without forgetting the previous naming. This famous mythic scene wherein God bequeaths the first human being with the power to bestow all living creatures with a conceptual tag is often discussed in classes leading to confirmation or *bat mitzvah* as a sign of human authority over animals, yet, in the realm of psychoneurology, naming is most commonly approached as a means of ordering the world, transforming the plenum into individual creatures and things of the world, that is, a means of creative selective seeing.

Listen to these two voices, one from religion and one from psycholinguistics. First, Rabbi Andrew Davids, in his article "The Power of a Name; The Power of Naming," states, "Just as God separates light from darkness and dry land from water, this biblical text affirms that humans... may seek to bring order to our chaotic and dynamic world through the process of naming." Second, Wendell V. Harris, in his article "Adam Naming the Animals: Language, Contexts and Meaning," warns us that "Language can no longer be seen as labeling discrete-objects-in-the-world but as dividing up the most primitive data presented to the mind... according to the human frame of reference."

Recall that the most often reported location of self is in the head. Adam's bestowal of names is, from the viewpoint of Nihilology, the epitome of how the eyes are used by the self located in the head: all animals (we will contend with things next) in the world have a name. Name is laid over the animal via our concepts, however, from the location of the head, all animals are seen as imbued with their name. In a culture of head-located selfhood, forgetting the name of an animal, or person, is one of the accepted signs of Alzheimer's disease or other form of dementia. So ingrained is it to see names as integral to the nature of things.

Now, to illustrate how the eyes see **things** when self is located in the head, let us move forward in time to the writings of Sen-No Rikyu, a Japanese tea master who lived in the sixteenth century. Most point to him as the originator of the way of tea in the simple style. It embodies the philosophy of *wabi-sabi* to inspire the aesthetic mood through the appreciation of the simple and rustic: light filtered through pines, movement of walking and kneeling, form of an unfinished gate, and a cracked tea bowl.

The garden path approaching the teahouse is called the *roji.* It has two sections, outer and inner, with gates along the way to provide thresholds to the path. The teahouse guests' slow walk from the outer garden through the inner garden to the porch of the teahouse is consciously designed to serve as a transition from the world of busyness to the world of *wabi-sabi:* the interior of the teahouse infused with the aesthetic mood.

Now, to shed light on our analysis of the objective world of objects as seen from the head-located self, listen to these verses from the master Sen-No Rikyu, as quoted in A.L. Sader's book *Cha-no-yu: The Japanese Tea Ceremony*:

While the *roji* is meant to be a passageway
Altogether outside this earthly life
How is it that people only contrive
To sprinkle it with the dust of the mind

Though I sweep and sweep
Everywhere my garden path
Though invisible
On the slim pine needles still
Specks of dust may yet be found

Here Sen-No Rikyu laments that, though the *roji* offers us the opportunity to transition away from worldly life, it is our ingrained bent to dirty up the path with our memories, concepts, and names. In Rikyu's poesy, memories, concepts, and names are called "dust of the mind." They need to be swept away by walking along the *roji*. And, for Rikyu, not a single speck of concept can remain in the aesthetic realm of the teahouse if the tea ceremony is to find its mood.

So, the *roji* is the transitional space between the world of daily associations and the aesthetic world of soft-lined objects, rich light, and beautiful movement, that is, the transition that wipes the mind free of concepts. Outside is the realm of objects laden with associations: cars, desks, digital devices. The transition occurs in the realm of nature, objects freed up of associations: trees, rocks, bamboo, moss. The whole point here is that when the self resides in the head, concepts dominate the objects perceived. The *roji* is consciously designed to dust off objects and make them clean of thinking.

The underlying axiom in this *wabi-sabi* philosophy is that all objects of the busyness world are suffused with memory/learning etc. These associations contribute to what is perceived as essential to the object. Yet, with this cleansing of associations, in the language of Four-Staged Kenosis with the descent of self location from the head to heart, the object is perceived to be quite a different thing altogether. Remember, psycho-neurology has taught us that seeing is a creative selective process, that "we re-construct the reality around us and we glean "paltry samples from the cornucopia of information that the world puts on offer."

Now, upon descending to the second stage in the kenosis process, a person enters into a sensate relationship with the world that Nihilology calls aesthetic. In philosophical circles, aesthetics is the study of the interaction of our senses with our emotions, the most important of the senses in this regard—and regarding the human sensate relationship with the world generally—is vision. This interaction of vision with emotion, according to the aestheticist, is termed "taste."

The Stanford Encyclopedia of Philosophy puts forth fertile words on what aesthetic scientists term "the immediacy thesis." The base notion is that judgments regarding beauty are not at all mediated by rational inference, principles of exegesis, or any other employment of concepts, "but rather have all the immediacy of straightforwardly sensory judgments... we do not reason to the conclusion that things are beautiful, but rather 'taste' that they are."

Until the middle of the twentieth century the study of taste had focused on objects of fine art with the goal of determining their unique characteristics that gave birth to the aesthetic mood in an onlooker. But in recent generations the study of taste has evolved into what is termed "everyday aesthetics," and the focus has grown from the objects of fine art to an expansive focus on the **interaction** of the onlooker with all objects of the world. Thus, there exists in the world no aesthetic objects *per se*, there are only aesthetic perceptions of everyday objects (and fine art objects) whose aesthetic qualities are tasted by the perceiver. With this evolution, aesthetic philosophy has come into accord with modern research in the fields of neuroscience, experimental aesthetics, and interaction design.

Everyday aesthetics found its birth in the 1934 book by John Dewey titled *Art as Experience*, often deemed the most important American work every written on the subject. In it he bemoans the effete trans-worldly status that fine art had taken on in the view of his educated peers who glorified fine art by setting it upon a rarely reached, high pedestal. "For many persons an aura of mingled awe and unreality encompasses the 'spiritual' and 'ideal' while 'matter' has become by contrast a term of depreciation, something to be explained away or apologized for."

Dewey proceeds with his populist philosophy toward grounding the aesthetic in the experience of the onlooker with the objects of the world, wresting it away from a particular etheric sort of object. In this emphasis upon the interaction between an onlooker and objects of looking, Dewey welcomed the things of everyday, and everyone's, life—pedestals, frames, walls, doorways—into the formerly elite club of the aesthetic. In so doing, Dewey also provided ground for the contention of Nihilology: when self is located in the heart, the seat of emotion, and the second stage of descending kenosis, the eyes are employed to experience the objects of the world as imbued with aesthetics.

With these religious, philosophical, and scientific notions in hand, drawn from the Hebrew Bible, the *wabi-sabi* of Japanese garden design, aesthetics, and neuropsychology, we conclude there is a direct correlation between the location of self and the way self employs the eyes resulting in particular modes of perception of objects in the world. Perhaps this heuristic mnemonic will be helpful to our readers:

When self is located in the head, we **conceive** the world and experience its objects imbued with associations.
When self is located in the heart, we **taste** the world and experience its objects imbued with aesthetics.

Now we will add one final stanza to our mnemonic:

When self is located in the stomach we **will** the world and experience its objects as imbued with dynamics.

If we look again at the graphic above, we see it posits an exact correlation among the third stage of the outsight model, the third stage of the kenosis model and perception of the objects of the world as imbued with dynamics. How is it Nihilology can make this correlation?

The third stage of the outsight model refers to the realm of the extra conceptual wherein, as we discussed earlier, a scientist, zealously attempting to solve a conundrum, moves out of the realm of thinking and begins to assail the conundrum unconsciously. To this unconscious problem-solving psychologists apply the term "parallel processing." Remember this quote from Anselm of Canterbury lamenting his struggles over an intellectual challenge:

But just when I wanted completely to exclude from myself this thinking
just then it began more and more to force itself insistently upon me unwilling and resisting [as I was]
Then one day when I was tired as a result of vigorously resisting its entreaties
what I had despaired of [finding] appeared in my strife torn mind
in such way that I eagerly embraced the [line-of-] thinking
which I, as one who was anxious, had been warding off

We used this quote earlier as an illustrative instance of how problem solving takes place outside of conscious awareness. But this quote also shows a classic Freudian description of unconscious forces thrusting themselves into consciousness in the midst of a conscious attempt to suppress those forces. Anselm was engaged in a wrestling match, a battle of his person opposing an impersonal might.

The impersonal quality of the unconscious is stressed in Freud's naming. "Id" is Latin for "It" which was originally presented as the German term "Es." And this "It" was posited to be a "thriving cauldron" of impulses epitomized by the impulse toward life (*Eros*) and the impulse toward death (*Thanatos*) in struggle for dominance.

In an article from the web journal *Psyche* titled "Your Hidden Unconscious Mind," we learn that Freud envisioned the unconscious as a storehouse set aside specifically for socially unacceptable desires, traumatic memories, and painful emotions, all well-guarded by a psychological mechanism termed "repression." "However, the contents did not necessarily have to be solely negative. In the psychoanalytic view, the unconscious is a force that can only be recognized by its effects—it expresses itself in the symptom."

So, for Freud the unconscious is a place of forces, dark to the person until they burst out into awareness in the form of psychological symptoms. This view of Freud's is spot on regarding the force-nature of the unconscious, and it correlates nicely with the third stage of the kenosis of self wherein the self is located in the stomach, that is, the force-nature of the unconscious jives perfectly with the will of the self located in the stomach. Both exhibit force as their essential nature.

However, Freud's perspective on the unconscious is limited in that it does not take into consideration the pattern of research that we have analyzed regarding the teachings of the world's religions on means of staying awake in the extra conceptual (the unconscious) and the realm of the super psyche. It also does not exhibit any cognizance of the narrative evidence available to him, such as that averred by August Kekulé in his "*träumerei*" discovery of the benzene ring. Further, Freud, of course, knew nothing of the twenty-first century psychoneurological research on the variously located self, and unfortunately, Freud's *heimwelt* of knowledge was not open to the Japanese teachings on the *hara*.

Recall earlier we quoted science writer Emma Marris who pointed out that the stomach is often experienced as a location of self. Marris went on to say that the stomach is "the point from which we initiate action." Now the science of biophysics tells us our body has a center of gravity that serves as its balance point or the point about which our body balances without a tendency to rotate. Also, when we are in a standing position, the center of gravity for the human body is in the region of the stomach. Further, the science of kinesiology tells us that movement of the body initiated from the stomach is easier and more powerful than movement initiated from parts of the body away from the center of gravity, such as the shoulders or the head.

These bodily sciences of the West, however, are rudimentary when viewed in light of the sophisticated physiological system of the *hara* developed in Japan over the past 1,500 years. In complete accord with western biophysics and kinesiology, there is alive in Japan a culture-wide axiom that *hara*, the belly, is the center of gravity in the human body and the source of force for physical movement. Indeed, *hara* is experienced as the source of all forces within the body.

Probing a bit deeper into the philosophies of Japan, we read that the martial arts do not allow for a force boundary at the skin. Indeed, they take as a ground experience that *hara* projects bodily force out into the world and in so doing influences the objects and events in the objective space. Now this perspective may strike our readers as a form of psychokinesis befitting the magician's stage, not the stage of objective world. However, this philosophical thought is simply a visceral expression akin to our common metaphorical language of "will power." The power of our will extends out into the world and accomplishes great tasks. The Japanese martial artist experiences will power as a force emanating from their stomach!

Here is another example of Japanese visceral philosophizing, this time from the article "About the Art of Aikido." The name "Aikido" can translate as the "Way of Harmonious Force." We suggest for our readers' sake that "Aikido" connotes a martial discipline that joins the forces of the body with the forces of the objective world. "Much like a hurricane, tornado or tidal wave, the forces found in nature are efficient, rational, and soft, while the center is immovable, firm, and stable... This principle of a firm center and a soft, adaptable periphery is universally consistent—and must be true for each person, as well."

Note here the internal physical forces "for each person" are likened to, even equated with, the forces of the objective natural world such as hurricanes or tornadoes.

For our discussion of the variously located self, the essential point in this discussion of *hara* and its parallels in the West is that the stomach is shown to be a juncture for the physical internal realm and the objective world.

Now, recall our earlier discussion of the stomach as a second brain. This brain in the stomach shows much of the multifaceted functioning as the brain in the head, including complex neural networks and the employment of neurotransmitters such as serotonin. But it must be remembered that the second brain is not the purveyor of concepts. Its great talent is one of force sensitivity. For example, the stomach "understands" threat and expresses this threat as butterflies, and if the threat is deemed to be intense, it cramps itself up. The stomach is the body's watcher of the forces coming at it from the surrounding objective world, and the first brain may be utterly deaf to its independent duties. As Emma Young, Editor of the *New Scientist* journal, states in her article "Gut Instincts. Secrets of Your Second Brain: "... although you are not conscious of your gut 'thinking,' the ENS [enteric nervous system] helps you sense environmental threats, and then influences your response... A lot of the information that the gut sends to the brain affects well-being and doesn't even come to consciousness."

This language of the stomach's purveyance of forces as a type of understanding is often taken up by westerners in an attempt to educate readers about this cultural axiom of *hara* in Japan. For instance, in an article titled "The Hara: Center of Being," Sharon Morelli states that *hara* is a Japanese term that denotes the middle of the human body, just behind the belly button. She then translates its meaning for the understanding of western readers, "This is said to be the intersection of mind and body, and the seat of all... 'gut' knowings. For some people, the *hara* is the primary source of information about the world."

So, to conclude, we propose a consistent correlation between the third stage of our outsight model, the third stage of our kenosis model, and a specific perception of the objects in the objective world. In the third stage of outsight our conundrum-seeking scientist has sunk through the stage of waning conceptual activity into this third stage typified by parallel processing and unconscious forces. Correlated with this sinking is a passing through self located in the "third brain" of the heart and residing in the second brain of the stomach (*hara*) as the seat of the physical force we call the will. And as a clear inference from all the above, Nihilology holds that when forces of the unconscious are dominant and the self is located in the seat of physical forces then the eyes will perceive the objects of the world as imbued with forces:

momentum, inertia, acceleration, and speed or, if the objects are human, threat.

So that our technical terminology is consistent, easily remembered and easily employed by our readers, we must explain that Nihilology uses the term "dynamics" to typify how a person whose self is located in the stomach perceives the world, whereas it might seem more straight forward to call on the term "forces." We do so because Nihilology's unique interpretation is that the very **objects** of the world are perceived by the eyes quite differently according to specific self location. Now the sort of mathematics that describes the interaction of physical forces with physical objects is called Dynamics. As Glenn Elert, in his ebook *The Physics Hypertextbook* blithely states, "When you run out of things, you run out of forces."

So, let us recall our three-parted heuristic mnemonic introduced before. When self is located in the head we conceive the world and experience its objects imbued with associations. When self is located in the heart we taste the world and experience its objects imbued with aesthetics. When self is located in the stomach we will the world and experience its objects as imbued with dynamics.

This interpretation by Nihilology should not be heard by readers as so far removed from our common experience. In your mind's eye envision a gathering of people gazing at the flag of the US furling in the wind. Most see the symbolic nature of a pennant: loyalty to the president or the history of the stars and stripes. Remember Adam bestowing names on the animals and the *roji* path cleansing objects of associations. Other people see the aesthetic quality of colors red, white, and blue, the array of the stars, the light coming through the cloth. Remember, in the studies of everyday aesthetics all objects can reveal their beauty according to the aesthetic mood of the viewer. And finally, some people see the dynamics of the flag: movement of the waving, the bending of the pole.

We hold that these flag-gazing people, with the aid of Nihilogical understanding and reflection on where their self is located, could be aware that they can use their eyes as tools, initiating this use by a self watching through the eyes from various locations in the body: head, heart, and stomach. Our perspective, however, does not rest there. Indeed, it is unique and powerful in that we take these differing perceptions to their logical and perhaps shocking conclusion: these perceptions, each correlated perfectly with a location of self, are actually revealing three distinct spatial dimensions of the objective world.

Furthermore, these dimensions of associations, aesthetics, and dynamics must be taken to satisfy, at least in part, the demand of string theory for extra spatial dimensions. We interject "in part" because string theory demands at least six. This is the unique interpretation of Nihilology.

Recall the words of Donald Hoffman from his TED Talk titled "Do We See Reality." We cited Hoffman earlier in this discussion, but with all the enhancing argument presented since, his outsight should have a deeper impact on readers, "Once we let go of our massively intuitive, but massively false assumption about the nature of reality... reality is more like a 3D desktop... Space as you perceive it is the desktop and physical objects are the icons in that desktop..."

This notion of Hoffman's is supported by wide-ranging research; it is a neurological fact that has aided us in our arguments regarding the extra dimensions demanded by string theory. Indeed, the objective world is an objective **space**, a stage upon which objects imbued with various qualities appear to human seeing: imbued with associations when the eyes are used by a self located in the head, imbued with aesthetics when the eyes are used by a self located in the heart, and imbued with dynamics when he eyes are used by a self located in the stomach.

Now, once again we ask in our readers' stead: are these three sorts of visual perception of the objective world really experiences of spatial dimensions just like width, length, and height? Let us look at three illustrative modes found throughout the POPs canon for explaining a dimension as empirically spatial and apply them to associations, aesthetics, and dynamics.

First, it is taken as an axiom among the POPs that humans observe three spatial dimensions that carry the names width, length, and height. This observability provides the empirical evidence for the reality of the three. We quoted Brian Greene, M. Seifert, and Lawrence Krauss as people who maintain this axiom. We went forth to argue that the perception of only three spatial dimensions is an intersubjective knowing rather than an objective fact, but nonetheless the observability of three spatial dimensions is held by the POPs to be probative of their factual nature.

Now, our arguments for the extra dimensions that we name associations, aesthetics, and dynamics have been based on research in the areas of philosophy, psychology, and neurology regarding the human sense of vision. When the self is located in the head and uses the eyes as a tool, the person sees objects with associations. When the self is located in the heart and uses the eyes as a tool, the person sees aesthetic objects. When self is located in the stomach and uses the eyes as a tool, the person sees dynamic objects. So, the perspective offered by Nihilology finds its origin in human vision. Thus, our dimensions fulfill the observability criterion put forth by the POPs for the reality of a spatial dimension.

Second, in order to explain the common three spatial dimensions and one of time, we routinely hear in the POPs canon the simple story of these dimensions as coordinates that allow people to meet. The story goes that two friends agree to come together at the corner of such street (X axis or width) and such street (Y axis or length) on a certain floor (Z axis or height) at a certain hour of the day.

Now it is our contention that the very same story can be employed to explain the dimensions of associations, aesthetics, and dynamics. Let's say two friends, each with self residing in the head (as we have learned is most often the case) agree to meet at the CN Tower in Toronto at the corner of Front Street West and John Street on the Glass Floor Level at 2:00 pm on Tuesday. Being aware that their variously locating self may be in a more rarefied stage of kenosis at the actual point of meeting, they further agree to meet, if need be, at 1500 lumens of light or 4 feet per second of speed.

Once again in our readers' stead, we ask: even if one friend were to arrive with self located in the heart and the other friend were to arrive with self located in the stomach, would they not still be able to meet in that the common width, length, height, and time do not disappear with the kenosis of self and the advent of the aesthetic dimension or the dynamic dimension? Would the friends not, regardless of self location, still find each other in the CN Tower on the Glass Floor Level at 2:00 pm?

Indeed, our friends would occupy the same three spatial dimensions of width, length, and height, and share the time dimension as well. However, locating their self in differing stages of kenosis, they would pass as ships in the night, or more technically as selves in distinct dimensions. One friend would be using their eyes from the self location of the stomach to perceive the objects of the world as dynamics. The Glass Floor Level would be an aggregation of objects moving about by force of muscle or electric wheelchair, each object with moving skins and moving mouths. Of course, none of these would have names or attached memories, and their sounds would carry no meaning. The other friend would be using their eyes from the self location in the heart and perceiving the world as aesthetics. They would experience the Glass Floor Level filled with objects of light exhibiting various brightness and color.

Though our friends might each recognize a hoard of objects milling about the observation deck, they could not recognize each other. "What was your name?" "Your position in the company?" They would have no conceptual tags to identify each other. One would be searching for an object with a certain intensity of light: 1500 lumens. The other would be seeking an object with a certain speed: 4 feet per second. Having arranged to meet while both sharing the self location in the head, they would fail to actually connect were they to arrive in the self location of heart and stomach. They would fail to coordinate.

And finally, the third mode found throughout the POPs canon for explaining spatial dimensions might be called the "crossing-over" that states: A force, particle or object of a certain nature in one dimension is experienced as having a different nature when it crosses over to another dimension.

The nineteenth-century novella by Edwin Abbott *Flatland: A Romance of Many Dimensions* seems to have been the precursor to all the POPs crossing-over illustrations. Abbott tells the delightful story wherein an apostolic sphere, existing in a three-dimensional scape called Spaceland, visits, on an edifying mission, the two-dimensional scape of Flatland where "he" appears as a circle.

We have entertained the crossing-over from many popularizers of science in earlier parts of our book.

Recall Theodor Kaluza, in seeking out the unity of gravity and electromagnetism, discovered that electric charge is equivalent to movement in another dimension and movement in this other dimension is equivalent to charge. Thus, electric charge in the three spatial dimensions (plus the fourth as time) can be understood as motion along the fifth dimension.

David Bohm, in his own universalist language, tells us that the plethora of quantum particles in our world of four dimensions are, quite literally, projections of a higher dimensional reality in that these particles cannot be understood solely via interactions among themselves.

Recall also, Lisa Randall explains that gravity is carried by a specific particle called the graviton which is the manifestation in the realm of three spatial dimensions of a vibration from a string in an extra dimension. Brian Greene chimes in here as well when he states that a string vibration in dimensions six—ten is empirically detected in the realm of four dimensions as a particular particle with its own mass.

This crossing-over explanation, as witnessed throughout the POPs canon, applies just the same to our proposed extra dimensions of associations, aesthetics, and dynamics. An object that is perceived as imbued with associations by a person whose self is located in the head is perceived as imbued with aesthetics by a person whose self is located in the heart. And that very same object is perceived as imbued with dynamics by a person whose self is located in the stomach.

Before concluding this phase of our discussion exploring extra dimensions as demanded by string theory, we might remind our readers that Nihilology does not posit an idealist's world. The three spatial dimensions of width, length, and height, as well as time, abide with the advent of the extra dimensions of associations, aesthetics, and dynamics. The objects of the world, the grounding of our experience of objectivity, also remain. Just as is the case with the forces and particles posited by Kaluza, Bohm, Randall, and Greene, the imbued nature of the objects of the world appear differently when they cross over into a different dimension.

So, with all this argument on our part, it seems clear that the popularizers of science must accept the dimensions of associations, aesthetics, and dynamics as empirical realities. They meet the criterion of observability, the criterion of coordinates, and the criterion of crossing-over. These are the very criteria used by the POPs themselves to establish the empirical reality of spatial dimensions.

Please note again we have not focused on the fourth stage of outsight or self location/kenosis in the previous argument because the fourth stage is beyond the end of the continuum in each model, the continuum of concepts in the case of the outsight model and the continuum of selfhood in the case of the four-staged Kenosis model. Thus, they are not directly explanatory in proposing extra spatial dimensions as demanded by string theory. As we stated earlier, the fourth stage in both models is below the continuum of objective experience. And to propose objective experience of extra spatial dimensions has been our aim in this discussion.

However, below the continuum of concepts and below the continuum of selfhood lies the realm of Nothing. Nothing serves as the culmination of the rarefying phases of these processes and also serves as our guide in analysis of these processes. Nothing is the ground upon which our proposition of extra dimensions is based. Remember our earlier discussion where we established it is Nothing that makes the world real, and it is Nothing that makes the world work!

Let us now enter a detailed analysis of the nature of visual perception of the objective world when the self is located in the heart and uses the eyes as a tool to experience the dimension of aesthetics. We choose to go into depth about the dimension of aesthetics for three reasons. First, recall that Adam bestowed names on the animals. In the language of Nihilology, objects of the world are **painted over** with concepts, memories, and ideologies in the dimension of associations. This seems to imply that objects are not inherently imbued with associations. While we hold that associations have the same relationship to the inherent nature of objects as exhibited by aesthetics and dynamics, it is a rather counter-intuitive notion, so we will offer analysis of the dimension of aesthetics as it is easily intuitive to see light, color and shape as inherent to objects and thus such an analysis will strike our readers as more impactful. Second, remember that most people, most of the time, perceive the objective world from a self located in the head, thus the visual experience of objects seemingly imbued with associations is a cultural *de facto* norm. Again, making our argument based on evidence regarding aesthetics, that is, a dimension that readers do not find commonplace, or "the fact," will have greater impact for readers. Third, the evidence for the perception of the aesthetic dimension is readily available from accounts of people with extensive scientific training, thus providing credibility for our argument with a wider range of readers.

Perhaps the finest documentation of perception of what Nihilology terms the spatial dimension of aesthetics is provided to us in the writings of Barbara Ehrenreich. Her words are precise and provocative, and in reflecting on her aesthetics experiences, she does much of our analytic work for us.

Ehrenreich earned an undergraduate degree in chemistry, followed by a Ph.D. in cellular immunology. All her young life she intended to be a scientist but took another trajectory in her professional career as a writer and activist. She is famed for her progressive political writings, crowned by her 2001 classic commentary on low-wage workers *Nickel and Dimed: On (Not) Getting By in America.* She taught as a professor for many years at universities around the US and is currently at SUNY Westbury. When called upon to describe herself, Ehrenreich proclaims she is a "professional myth buster by trade" and a "fourth generation atheist."

The circumstances that brought Ehrenreich forward into the public arena with her aesthetics encounter are quite dramatic. They point to the cultural barrier erected by the popularizers of science that deter their own colleagues from writing of their entrances into the dimensions of associations, aesthetics, or dynamics, the very dimensions that POPs seek as a result of the demand of string theory.

Just at the turn of the century, Ehrenreich was faced with two tough challenges. First, she was tasked with gathering up a lifetime of jots, notes, and unfinished manuscripts for transfer to a controlled atmosphere library for safe keeping so they would be available for future scholars to explore. At the same time, she was undergoing surgery and chemotherapy for breast cancer. These two in cahoots, sending her writing life to storage and contemplating the possibility of an early physical death, created in her the gumption to publish her experiences of what Nihilology calls the dimension of aesthetics. Prior to that time, she had kept her manuscript that analyzes these encounters to herself, mainly out of fear that her colleagues would deem her mentally ill for having the experiences or be intellectually offended at her analysis.

Early on in Ehrenreich's adulthood she had an extensive period of residence in what Nihilology calls the spatial dimension of aesthetics. Recall from our discussions before that, for human beings, the dominant sense that links us to the objective world is vision, and the dominant quality of objects in the aesthetic dimension is light. Ehrenreich perceived visually, in a rather dramatic fashion, this quality of light. And as she returned in later life to write about this dimension, she maintained a rigorous scientific analytical perspective. Upon looking into her own encounters with the aesthetic, she realized many others had also done so, and they had also found ways of talking about them in public. But she found their vocabulary of "mystical experience" to be utterly repulsive. "As far as I was concerned—as a rationalist, an atheist, a scientist by training—this was the realm of gods and fairies and of no use in the human project..."

The analytical manuscript that Ehrenreich decided to bring into the popular view resulted in her 2014 book called *Living with a Wild God: A Non-Believer's Search for the Truth About Everything* which will serve as the root source for this discussion. She reflects thus on its title, "Why would I want to apply the ancient, well-worn notion of "God" to that... which bore not the slightest resemblance to anything in the religious iconography.... There had been no soulful, long-suffering face, no accompanying cherubs or swooning Madonna—no face at all, in fact."

Indeed, we ask the same question of her. Why did Ehrenreich apply the term "God" in her book title to the light she encountered while arguing within the text that this light had "not the slightest resemblance to anything in the religious iconography." We can only surmise that Ehrenreich employed the same stratagem we talked about earlier in the case of book titles by such POPs as Lisa Randall, Carl Sagan, Michio Kaku, Lawrence Krauss, and Frank Wilczek. She calls on the evocative power of a religious book title to draw in readers and at the same time upholds her scientific atheism within the book itself.

Allowing for this critique as an aside, let us investigate the specifics of Ehrenreich's excellent analysis. First, recall that POPs present as objective fact that human beings observe only three spatial dimensions whereas we propose that extra spatial dimensions can be observed—with the physical eyes—according to the location of the self. Nihilology suggests that this position of the POPs is not an objective knowledge, rather an intersubjective knowing, one verified not in the objective world but in the world of relationships among a group who agree. Ehrenreich goes straight to the heart of this matter in full accord with our view, "Human beings are connected... by our joint agreement about the 'real'... which I was expected to contribute to as a scientist... Step outside the borders of what is 'real' and ... you might as well fall right off the planet into a personal orbit of your own."

Ehrenreich is fully aware that entering the perception of a world of objects imbued with light puts her just outside what popularizers of science collectively agree upon as real. Her trepidation at becoming an alien among her science peers kept her on their intersubjective planet for much of her professional life. Yet, with the inspiration of her scholarly life retrospection and her cancer, she gained the gumption to aver that her perception of the realm of light is well within the sphere of empirical knowledge, not originating in any sort of faith. According to Ehrenreich, the empirical sphere is simply a bit wider and more dimensional than her common colleagues allow, and she invites them to open their minds and eyes to this more dimensional empirical. We join her in this invitation.

Ehrenreich also stands with Nihilology in averring that the extra dimensions, in this case the dimension of aesthetics, are material spatial dimensions, objectively existent alongside and permeated by the daily dimensions of width, length, and height. Ehrenreich maintains that her perceptions occurred on the stage of the physical, as witnessed by her account of maintaining bodily control and her paean to three-dimensionality. On one particular day she describes walking toward the sky that was bright with light, being astounded that she was moving under her bodily exertion through three-dimensional space, yet pulled to the light, "During my experience... nothing unnatural or physically impossible had occurred: objects had not moved on their own, and the laws of geometry had remained in force."

Going a bit deeper in our analysis, we point out that Ehrenreich's experience in the dimension of aesthetics affirms and correlates nicely with three central aspects of our dovetailed models of outsight, awe and Four-staged Kenosis. First, it reveals a waning of conceptual activity as called for in stage two of our outsight model. Second, Ehrenreich's experience of ecstasy correlates with location of self in the heart and the arising of awe. And third, it gives witness to an ebbing of selfhood, i.e., what we have discussed as a rarefying of self, as called for in the Four-staged Kenosis model.

First, Ehrenreich's analysis of the dimension of light repeatedly states that she lost the ability to perform cognitive functions therein, "Here we leave the jurisdiction of language... when you take away all human attributions—the words, the names of species, the wisps of remembered tree-related poetry, the fables of photosynthesis and capillary action—*there is still something left.*"

We have accounted for this loss of cognition in our outsight model as a shift from left brain to right brain activity and within our Four-staged Kenosis model as a descent of self location from the head to the heart. Furthermore, and important to our discussion of perception of the objective dimension of aesthetics, Ehrenreich not only is aware of her inability to conceive of things, but also that those very things are stripped of attributions, "the names of species" are taken away. In the same discussion, she calls the dimension of light "a world drained of referents and connotations." She holds that the **objects of**

the world are drained of connotations, not "my head" was drained or "I" was drained. This perspective is fully in line with the interpretation of Nihilology that the objects of the world become painted over with associations when one is thinking with self located in the head and that objects are cleansed of these when one moves to right brain activity and locates the self in the heart. Yet there is "still something left." In this dimension what is left is objects imbued with aesthetics.

Secondly, Ehrenreich lavishes her readers with elegant descriptions of the ecstatic tone during her time in the dimension of aesthetics. Recall, according to our model, awe comes to bloom with the simultaneous advent of three characteristics: waning sense of self, ebbing of conceptual activity and presentation of a visually vast field. The ecstasy that Ehrenreich reports, we would argue, is equivalent to awe when the person is not presented with a perceptual space that is vast. Also of note is that Ehrenreich differentiates between ecstasy and emotion, "'Ecstasy' would be a word for this, but only if you are willing to acknowledge that ecstasy does not occupy the same spectrum as happiness or euphoria, that it participates in the anguish of loss..."

Earlier we presented a history of the philosophical musings and the modern-day neurological research on awe. We learned that awe is an utterly impactful experience, yet it is not solely positive for the experiencer. It includes fear. This is echoed in Ehrenreich's ecstasy that "participates in the anguish of loss." Nihilology interprets this as anguish at the loss of accustomed self location in the head, as the anxiety that accompanies the rarefying of self and taking up residence in the heart. Ecstasy is not simply a deeper or subtler emotion; it is another sort of experience as found in the dimension of aesthetics.

Thirdly and finally, Ehrenreich's analysis of the dimension of aesthetics gives unambiguous witness to the rarefying of self as called for in our

Four-staged Kenosis model. Behold her view of light, identified here as "the blaze," as she was out on a pre-dawn stroll. No sacred visions appeared, no deep voices from a dramatic sky, no symbolic totem animals. "You cannot observe fire really closely without becoming part of it. Whether you start as a twig or a gorgeous tapestry, you will be recruited into the flame and made indistinguishable from the rest of the blaze."

Recall how, at a preliminary culminating point in our discussion, we stated that Nothing is a declarative with gravitas, a declarative with a pull to it. And now that all the argument and narrative has ensued, we can reveal to our readers that this pull contains both an intellectual and experiential aspect to it.

Due to nature of Nothing as inhabiting a realm over the edge—beyond the continua of thinking, selfhood and materiality—the concept of Nothing has no pith to it. Nothing, appropriately so, is difficult to wield and is impossible to convey in a phrase or two. However, if readers persevere (and we are so thankful to those who stayed with us for 197 pages) they will have gained a wide appreciation of Nothing and its heuristic power in analyzing works of science. Remember that Nothing is not manifest in the world, yet it makes the manifest world work. Nothing is the empty hub that allows the wheel to spin. It is similar with the **concept** of Nothing; readers cannot lay an easy finger on Nothing, but it is a sharp interpretive tool. We have strived diligently to be an example of this heuristic and, to go one final step, we have strived to convince POPs that the appreciation of Nothing can be put to work in evolving their scientific understanding of the world.

The experiential aspect of the gravitas, the pull, of Nothing is what we feel in Ehrenreich's chronicle of the blaze. As she states, whether one enters the aesthetic dimension as a simple person ("twig") or as a sophisticated intellectual or artist (tapestry) one will be pulled toward the light and "recruited" to give up one's distance from the light. Recall our earlier discussion of Nihilitive Knowing where we concluded that Nothing cannot be experienced, that the rarefying process of self and other culminates in the greeting of Nothing with Nothing. We coined the term "inperience" to more accurately express this greeting.

It is this pull of Nothing that we feel from Ehrenreich. In entering the dimension of aesthetics, she has moved from stage one to stage two in the process of Four-staged Kenosis. Her self has sunk from the head to the heart. Making use of her eyes from this location in the heart, she perceives all objects of the world as imbued with light. She perceives light, but with this second step toward the fourth step of inperience, Ehrenreich feels the pull of Nothing, now just two steps away. She perceives the pull as originating from objects imbued with light, however these very same material objects in the world will be perceived imbued with forces if her self sinks another stage lower into the stomach, into the spatial dimension of dynamics. At that stage it will feel like the objects imbued with forces are pulling her to relinquish distance between them. And in the final stage of kenosis, she will feel the pull of objects imbued with Nothing, to relinquish distance between them and give rise to inperience.

Before concluding, let us remind our readers: just as Nothing lies over the cliff edge of thinking, over the cliff edge of selfhood, over the cliff edge of materiality, so also does Nothing lie over the cliff edge of dimensionality. Nothing is not another dimension; it is beyond dimension. However, Nothing gives reality to the realm of dimension. Were the POPs to study Nothing and come to an appreciation of the concept of Nothing, it would foster them to stretch toward Nothing and in so doing it would foster their understanding of extra dimensions.

In light of this long analysis, how will our proposal of extra spatial dimensions aid in string theory research? Let us note once more that Nihilology holds that the extra dimensions of associations, aesthetics, and dynamics are material spatial dimensions, just as is the case with width, length, and height. Making use of the physical eyes as tools of perception by a self in the position of head, chest, or stomach is backed by neuroscience and neuropsychology. And the perceptions of these tools of perception takes place in the realm of objective space.

We suggest that popularizers of string theory must become more informed in, and agile with, the research of psychoneurology in order to widen their notions regarding how human seeing actually takes place. As a result, they must surrender their intersubjective notion as to the un-observability of extra spatial dimensions which leads to the delusive conclusion that extra dimensions must be compactified. In the end POPs must look at the daily objective world as a conundrum to be explored. Mystery does not just lie at the small end of the continuum of materiality. The Large Hadron Collider is not the answer to all research into extra dimensions.

APPENDIX
Other Possible Realms

CASE 1: One-Dimensional motion of two objects in Two Dimensional Plane (w/Static Observer)

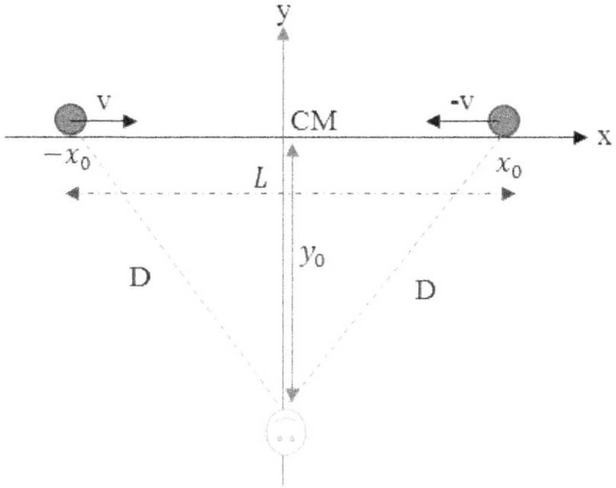

Initial conditions:

$L = 2x_0 \qquad D = \sqrt{y_0^2 + \frac{L^2}{4}}$

CASE 2: Observer moving through Three-Dimensional Space (w/Static Objects)

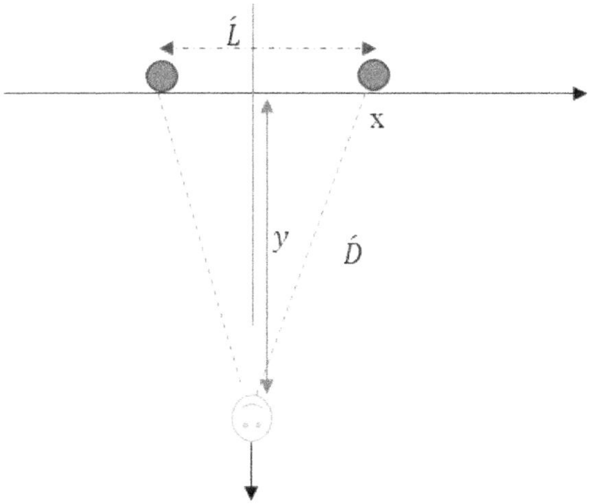

Apparent object motion in time t

$$\acute{L} = L - 2vt \qquad L + 2D = \acute{L} + \acute{D} \qquad \acute{D} = D + vt$$

$$y^2 = \acute{D}^2 - \frac{\acute{L}^2}{4} = (D + vt)^2 - \frac{1}{4}(L - vt)^2 = y_0^2 + 2Dvt + Lvt$$

$$position\ of\ observer \rightarrow y = \sqrt{y_0^2 + 2Dvt + Lvt}$$

$$velocity\ of\ observer \rightarrow u = \frac{dy}{dt} = \frac{vD + 1/2Lv}{\sqrt{y_0^2 + 2Dvt + Lvt}} = \frac{vD + 1/2Lv}{y}$$

$$acceleration\ of\ observer \rightarrow a = \frac{du}{dt} = \frac{(vD + 1/2Lv)^2}{y_0^2 + 2Dvt + Lvt} = \frac{(vD + 1/2Lv)^2}{y^2}$$

It may very well be that spatial dimensions beyond the three we have common experience of, are more easily understood by letting go of what the mind's eye can perceive and literally trust our gut as it moves through the higher dimensional space.

References

Abbott, Edwin. Flatland: A Romance of Many Dimensions. Mineola, New York. Dover Publications. 1992.

"About the Art of Aikido." Castle Rock Aikido. http://www.craikido.com/About Aikido.html

Aitchison, I. J. R.. "The Vacuum and Unification" in Saunders, Simon. The Philosophy of Vacuum. Oxford. Oxford Univesity Press. 1991.

Albert, David Z. "What is the Role of a Philosopher of Science." BigThink. January 4, 2010. https://bigthink.com/videos/big-think-interview-with-david-albert

Al-Ghazali: The Alchemy of Happiness. Claud Field (trans.). http://majalla.org/books/2005/The-Alchemy-of-Happiness.pdf

Al-Ghazali. "The Remembrance of Death and the Afterlife." (Book 10 of The Revival of Religious Sciences). http://lyconpolis.weebly.com/publisher/remembrance-of-death-and-the-afterlife-imam-ghazalipdf1

Allies, Mary H. (trans.) Defense Against Those Who Oppose Holy Images by St John of Damascus. https://www.christianhistoryinstitute.org/study/module/john-of-damascus

Allport, Gordon. The Individual and His Religion. New York. Macmillan Press. 1950.

Ames, Roger T. & Rosemont, Henry, Jr. (trans) Analects of Confucius. Nw York. Ballantine Books. 1999.

Anselm of Canterbury. (Jasper Hopkins trans.) Complete Philosophical and Theological Treatises of Anselm of Canterbury. Minneapolis. Arthur J. Banning Press. 2000.

Armstrong, Karen: In the Beginning. New York. Knopf. 1996.

Arango-Muñoz, Santiago."The Nature of Epistemic Feelings." Philosophical Psychology. (Volume 27, 2014). https://www.tandfonline.com/doi/abs/10.1080/09515089.2012.732002

Battersby, Stephen. 'It's Confirmed: Matter is merely vacuum fluctuations." New Scientist Daily Newsletter. https://www.newscientist.com/article/dn16095-its-confirmed-matter-is-merely-vacuum-fluctuations/

ben Jochai, Rabbi Simeon: The Sepher Ha-Zohar Or The Book of Light. (Nurho de Manhar, trans). https://wikischool.org/book/zohar

ben Judah he-Asid, David. Book of Mirrors/Sefer Mar'Ot Ha-Zove'Ot. Atlanta. Scholars Press. 1981.

Bergson, Henri. Matter and Memory. (Nancy Margaret Paul and W. Scott Palmer trans.) London: George Allen and Unwin. 1911.

Bergson, Henri. Creative Evolution. New York. Henry Hold and Company. 1911.

Blyth, R.H. Zen and Zen Classics. Volume Four. Mumonkan. Tokyo. Hokuseido Press. 1966.

Boaz, David Paul. "Shankara's Advaita Vedanta."

http://davidpaulboaz.org/_documents/stromata/shankaras_advaita_vedanta.pdf

Boesel, Chris & Keller, Catherine (eds.). Apophatic Bodies: Negative Theology, Incarnation, and Relationality. New York, Fordham University Press. 2009.

Bohm. David. Wholeness and the Implicate Order. New York. Rutledge. 1980.

Bohr, Niels. Atomic physics and human knowledge. New York. Chapman & Hall. 1958.

Bostrom, Nick. "Are You Living in a Computer Simulation?" Philosophical Quarterly (Vol 53. No. 211). 2003.

Bostrom, Nick. Superintelligence: Paths, Dangers, Strategies. Oxford. Oxford University Press. 2015

Bostrom, Nick. "The Simulation Argument." Future of Humanity Institute. February 21, 2013. YouTubre. https://www.youtube.com/watch?v=nnl6nY8YKHs

Boutroux, Emile. Pascal. Manchester: Sherratt and Hughes. 1902.

Bracken, Joseph. Subjectivity, Objectivity, and Intersubjectivity: A New Paradigm for Religion and Science. West Conshohocken, Pennsylvania. Templeton Foundation Press. 2009.

Brown, Frederick. "The Last Days of Jean Paul Sartre". The New Criterion. (September 1982). http://www.newcriterion.com/articles.cfm/The-last-days-of-Jean-Paul-Sartre-6523

Burke, Edmund. A Philosophical Enquiry About Our Ideas of the Origins of the Sublime and Beauty. Notre Dame, Indiana. University of Notre Dame Press. 1968.

Candia, Gustavo. "Causation by Nothing." (A Paper Submitted at the Institute of History and Philosophy of Science and Technology). 2005.

Carroll, Sean. The Big Picture: On the Origins of Life, Meaning and the Universe Itself. New York. Penguin Random House. 2016.

Carter, Brandon. "Large number coincidences and the anthropic principle in cosmology."

(Proceedings of the Symposium, Krakow, Poland, September 10-12, 1973). Dordrecht. Dordrecht. D. Reidel Publishing Co. 1974.

Chalmers, David. "David Chalmers & John Horgan." Bloggingheads.tv. May 22, 2007. https://bloggingheads.tv/videos/1187

Chen, Ellen Marie. "Nothingness and the Mother Principal in Early Taosim." International Philosophical Quarterly. (Volume 9. Issue 3).

CIN (Catholic Information Network). An Exact Exposition of the Orthodox Faith. https://www.ewtn.com/library/SOURCES/DAMASCUS.TXT

Clegg, Brian. A Brief History of Infinity: The Quest to Think the Unthinkable. London. Robinson. 2003.

Cohelho, Ivo. Hermeneutics and Method: The 'Universal Viewpoint' in Bernard Lonergan. Toronto, University of Toronto Press. 2001.

Cole, K.C. The Hole in the Universe: How Scientists Peered Over the Edge of Emptiness and Found Everything. New York: Harcourt, 2001.

Conway, Timothy. "John Scottus Eriugena: Greatest Theologian of Early Medieval Christianity." Enlightened-Spirituality.org. 2009. http://www.enlightened spirituality.org/John_Scottus_Eriugena.html

Copernicus, Nicolaus. On the Revolutions of the Celestial Spheres. World Digital Library. https://www.wdl.org/en/item/3164/

Copleston, F. A History of Philosophy. (Bk. I, vol. II). New York: Image Books/Doubleday. 1985.

Crick, Francis. The Astonishing Hypothesis: The Scientific Search for the Soul. New York. Touchstone. 1994.

Crick, Francis. What Mad Pursuit: A Personal View of Scientific Discovery. New York. Basic Books. 1988.

Davids, Rabbi Andrew. "The Power of a Name; The Power of Naming." Jewish Leaning. https://www.myjewishlearning.com/article/the-power-of-a-name-the-power-of-naming/

Davies, Paul. "Big Pictures of God." Closer to the Truth. January 28, 2016. YouTube.

https://www.youtube.com/watch?v=cp_7WSVmEZM

Davies, Paul. The Mind of God: Basis for a Rational Universe. New York. Simon & Schuster. 1992.

Dawkins, Richard. The God Delusion. New York. Houghton Mifflin. 2006.

Deane, Andrew R. "The Tea Garden." A Japanese Garden Handbook. https://japanesegardening.org/handbook/the-tea-garden/

de Beauvoir, Simone. Force of Circumstance. New York. G. P. Putnam's. 1965.

Dewey, John. Art as Experience. New York. Perigree Books. 1980.

Eckhart, Meister. Meister Eckhart: The Essential Sermons, Commentaries, Treatise, and Defense. Edmund Colledge (trans.). Mahwah, NJ. Paulist Press. 1981.

Eckhart, Meister. Meister Eckhart's Sermons. Christian Classics Ethereal Library. http://www.ccel.org/ccel/eckhart/sermons.ix.html

Eckhart, Meister. "Meister Eckhart (1260-1328) – Nondual Christian Mystic Sage." Enlightened-Spirituality.org. http://www.enlightened-spirituality.org/Meister_Eckhart.html

Editors. B'Or Ha'Torah: "Neurology and the Soul." Science, the Arts, and Problems of Modern Life in the Light of the Torah. (Number 10). 1998. https://www.simpletoremember.com/articles/a/neurology-and-the-soul/

Editors. "Ego. Superego, and Id." New World Encyclopedia. http://www.newworldencyclopedia.org/entry/Ego,_superego,_and_id

Editors. "Heisenberg states the Uncertainty Principle, 1927". A Science Odyssey: People and Discoveries. http://www.pbs.org/wgbh/aso/databank/entries/dp27un.html

Editors. "The Francis Crick Papers: The Discovery of the Double Helix, 1951-1953." Profiles in Science. National Library of Medicine. https://profiles.nlm.nih.gov/SC/Views/Exhibit/narrative/doublehelix.html

Editors. "Your Hidden Unconscious Mind." Journal Psyche. http://journalpsyche.org/your-hidden-unconscious-mind/

Edinger, Edward: Ego and Archetype. Boston. Shambhala. 1972.

Ehrenreich, Barbara. Living with a Wild God: A Non-Believer's Search for the Truth About Everything. New York. Hatchette Book Group. 2014.

Ehrenreich, Barbara. Nickel and Dimed: On (Not) Getting By in America. NewYork. Henry Holt. 2001.

Einstein, Albert. Relativity: The Special and General Theory. London. Methuen & Co. 1920.

Einstein, Albert. Living Philosophies. New York. Simon and Schuster. 1931

Elert, Glenn. "Forces." The Physics Hpertextbook. https://physics.info/newton-first/

Eriugena, John Scotus. Patralogia Latina. (Volume 122). latina.patristica.net

Evola, Julius. "The Japanese Hara Theory and Its Relations to East and West." Counter Currents. https://www.counter-currents.com/2013/06/the-japanese-hara-theory-and-its-relations-to-east-and-west/

Ferguson, Kitty. Steven Hawking: An Unfettered Mind. New York. Palgrave Macmillan. 2012.

Fetterman, Adam K & Robinson, Michael D. "Do You Use Your Head or Follow Your Heart? Self-Location Predicts Personality, Emotion, Decision-Making, and Performance." Journal of Personal Social Psychology. (August). 2013. https://www.ncbi.nlm.nih.gov/pmc/articles/PMC3722275/

Feynman, Richard. Surely, You Are Joking Mr. Feynman. New York. Norton. 1985.

Freud, Sigmund. Five Lectures on Psycho-Analysis. New York. Norton. 1990.

Garb, Johnathan. Shamanic Trance in Modern Kaballah. Chicago. University of Chicago Press. 2011.

Gaskill, J.B. "On Approach: Reflections on the G-d Discovery." The Policy Think Site. http://jaygaskill.com/OnApproach.htm

Gates, Sylvester James. "Superstring Theory: The DNA of Reality." The Great Courses: Physics Beyond the Edge. Teaching Company (video). 2006.

Genz, Henning. Nothingness: The Science of Empty Space. New York. Basic Books. 1999.

Gershon. Michael. The Second Brain. New York. Harper-Collins. 1999.

Ginsburgh, Rabbi Yitzchak. "Kaballah Concepts: Bitul." Mystical Paths. http://www.mpaths.com/2005/08/kabbalah-concepts-bitul.html

Gleick, James. The Information: A History, a Theory, a Flood. London. Fourth Estate, 2011.

Gleiser, Marcelo. A Tear at the Edge of Creation. New York. Free Press. 2010.

Gleiser, Marcelo. "Avoiding the Void: A Brief History of Nothing(ness)." Cosmos & Culture. November 2010.

Gleiser, Marcelo. The Island of Knowledge: The Limits of Science and the Search for Meaning. New York. Basis Books. 2014.

Goldman, David. "The God of the Mathematicians: The Religious Beliefs That Guided Kurt Gödel's Revolutionary Ideas." First Things. August 2010.

Goodman, Micah. Maimonides and the Book that Changed Judaism: Secrets of the Guide for the Perplexed. Lincoln. University of Nebraska Press. 2015.

Graham, Wallas. The Art of Thought. Kent, England. Solis Press. 1926.

Greene, Brian. The Elegant Universe: Superstrings, Hidden Dimensions, and the Quest for the Ultimate Theory. New York. Norton. 2003.

Greene, Brian. The Fabric of the Cosmos. New York. Knopf. 2004.Greene, Brian. "An Interview with Brian Greene." The Oxonian Review. April 4, 2011. http://www.oxonianreview.org/wp/an-interview-with-brian-greene/

Greater Good Science Center: https://ggsc.berkeley.edu/who_we_are

Guth, Alan The Inflationary Universe: Quest for a New Theory of Cosmic Origins. Reading, MA. Addison-Wesley. 1997.

Hajek, Alan. "Philosophical Heuristics and Philosophical Creativity." In Paul, Elliot Samuel and Scott, Barry Kaufman. The Philosophy of Creativity: New Essays. Oxford, Oxford University Press. 2014.

Harris, Wendell V. "Adam Naming the Animals: Language, Contexts and Meaning." The Kenyon Review. (Vol. 8, 1). 1986.

Hart, Ray L. God Being Nothing: Toward a Theogony. Chicago. University of Chicago Press. 2016.

Hayman, Ronald. Sartre: A Life. New York. Simon and Schuster. 1987.

Hawking, Stephen. A Brief History of Time. New York. Bantam Books. 1998.

Hawking, Stephen. The Grand Design. New York. Bantam Books. 2010.

Hawking, Stephen. "Stephan Hawking Makes It Clear: There is no God." CNet News. https://www.cnet.com/news/stephen-hawking-makes-it-clear-there-is-no-god/

Hawking, Stephen. "Stephen Hawking's Grand Design: Key to the Cosmos." The Meaning of Life. September 16, 2015. YouTube. https://www.youtube.com/watch?v=usdqCexPQww

Hawking, Stephen. "Stephen Hawking Talks About Reality." November 12, 2014. YouTube. https://www.youtube.com/watch?v=NUX6pfXRCLA

Heisenberg, Werner. Physics and Philosophy: The Revolution in Modern Science. New York. Penguin Books. 1962.

Hill, Kyle. "Thinking Scientifically." Science Based Life: Add a Little Reason to Your Day. 2012. https://sciencebasedlife.wordpress.com/2011/09/22/thinking-scientifically/

Hoffman, Donald D. "Do We See Reality?" TED: Ideas Worth Spreading. https://www.ted.com/talks/donald_hoffman_do_we_see_reality_as_it_is

Hoffman, Donald D. Visual Intelligence: How We Create What We See. New York. Norton. 2000.

Hofstadter, Douglas R. Gödel, Escher. Bach. New York. Basic Books. 20th Anniversary Edition. 1999.

Horgan, John. "Physicist George Ellis Knocks Physicists for Knocking Philosophy, Falsification, Free Will." Scientific American. July, 2014.

Horgan, John. Rational Mysticism: Dispatches from the Border Between Science and Spirituality.New York. Houghton Mifflin Co. 2003.

Horgan, John. The End of Science: Facing the Limits of Knowledge in the Twilight of the Scientific Age. New York. Perseus Books Group. 2015.

Hossain, Gahangir et al. "Understanding Effects of Cognitive Load from Pupillary Responses Using Hilbert Analytic Phase." Computer Vision Foundation. https://www.cv-foundation.org/openaccess/content_cvpr_workshops_2014/W09/papers/Hossain_Understanding_Effects_of_2014_CVPR_paper.pdf

Hough, Adrian. The Flaw in the Universe: Natural Disaster and Human Sin. London. John Hunt Publishing. 2010.

Husserl, Edmund. Cartesean Meditations. The Hague, Netherlands. Kluwar Academic Publishers. 1950.

Husserl, Edmund. Husserliana (Vol. XV). Catholic University of Leuven. Husserl Archives. 1973.

Hussey, Edward (trans.). Aristotle Physics: Books III & IV. Oxford. Clarendon Press. 1983.

Ibn Arabi. The Seals of Wisdom. Santa Barbara, California. Concord Grove Press. 1983.

Ibn Arabi. "The Book of Annihilation in Contemplation." The Muhyiddin Ibn 'Arabi Society. http://www.ibnarabisociety.org/articles/kitabalfana.html

Ippoliti, Emiliano (editor). Heuristic Reasoning. Heidelberg. Springer International Publishing. 2015.

Iraneaus. Against Heresies (2.10.3-4) http://www.orthodoxebooks.org/node/244.

James, William. The Varieties of Religious Experience. London. Penguin Books. 1982.

Jammer, Max. Einstein and Religion: Physics and Theology. Princeton. Princeton University Press. 1999.

Jaynes, Julian. The Origin of Consciousness in the Breakdown of the Bicameral Mind. New York, Houghton Mifflin. 1976.

John of Damascus. On the Divine Images. Crestwood, New York. St. Vladimir's Seminary Press. 1980.

Johnson, Claes. Dr. Faustus of Physics. Icarus eBooks. 2011. http://claesjohnson.blogspot.com

Jones-Smith, Elsie. Theories of Counseling and Psychotherapy: An Integrative Approach. Los Angeles. Sage Publications. 2016.

Joseph, Rhawn. The Right Brain and the Unconscious: Discovering the Stranger Within. Cambridge. Perseus Publishing. 1992.

Jurovics, Rabbi Raachel. "Believe in No-thing." Yavneh: A Jewish Renewal Community. http://www.yavneh-raleigh.org/teachings-from-the-rabbi/believe-in-no-thing

Kagge, Erling. Silence in the Age of Noise. New York. Pantheon Books. 2017.

Kaku, Michio. "It's Not All About You". Psychology Today. March 8, 2016.

Kaku, Michio. Physics of the Impossible. New York. Anchor Books. 2008.

Kaku, Michio. "Shocker: The Universe was Created." sciencevibe.com. YouTube. https://www.youtube.com/watch?v=0NbBjNiw4tk

Kaku, Michio. Visions: How Science will Revolutionize the 21st Century. New York. Anchor Books. 1998.

Kant, Immanuel. Critique of Judgment. J. H. Benard (trans.) https://ebooks.adelaide.edu.au/k/kant/immanuel/k16ju/complete.html

Kaplan, Robert. The Nothing that is: A Natural History of Zero. Oxford. Oxford University Press. 2000.

Kaufmann, Walter. Existentialism: From Dostoevsky to Sartre. New York. Meridian Books. 1956.

Keats, John. Ode on a Grecian Urn and Other Poems. Whitefish, Montana. Kessinger Publishing. 2010.

Kekulé, F. A. "Benzole fest: Rede." Berichte der Deutschen Chemischen Gesellschaft (23). 1890.

Keltner, Dacher. "Why Awe is Such an Important Emotion." Greater Good Magazine. (Video)

http://greatergood.berkeley.edu/gg_live/the_art_science_of_awe/speaker/dacher_keltner/why_awe_such_important_emotion/

Klein, Gary. Seeing What Others Don't: The Remarkable Ways We Gain Insights. New York. Public Affairs. 2013.

Koestler, Arthur. The Act of Creation. London. Hutchinson and Co. 1964.

Kohn, Livia. Sitting in Oblivion: The Heart of Daoist Meditation. Three Pines Press. 2010.

Krauss, Lawrence. A Universe from Nothing: Why there is Something Rather than Nothing. New York. Free Press. 2012.

Krauss, Lawrence. "Hidden Realities: The Greatest Story Ever Told . . . So Far." Conway Hall. December 16, 2016. YouTube. https://www.youtube.com/watch?v=BRNtcj6YRuc

Krauss, Lawrence. The Greatest Story Ever Told – So Far. New York. Atria Publishing. 2017.

Krauss, Lawrence. "The Greatest Story Ever Told . . . So Far." Talks at Google. June 5, 2017. YouTube. https://www.youtube.com/watch?v=x0mlkOZQ_fE

Kraut, Richard, "Plato", The Stanford Encyclopedia of Philosophy (Fall 2017 Edition), Edward N. Zalta (ed.). https://plato.stanford.edu/archives/fall2017/entries/plato

Kuipers, Ronald A. Critical Faith. Amsterdam. Rodopi. 2002.

Kurzweil, Ray. The Age of Spiritual Machines. New York. Penguin Books. 1999.

Larson, Gerald. Classical Samkhya. Delhi. Motilal Banarsidass. 1998.

Leane, Elizabeth. Reading Popular Physics. Farmham, England. Ashgate. 2007.

Lehar, Steven. "Toward a new ontology of brain dynamics: neural resonance + neuroacoustics."

Modern Dragons. http://moderndragons.blogspot.com/2009/11/toward-new-ontology-of-brain-dynamics.html

Lightman, Alan. "My Own Personal Nothingness." Nautilus. August, 2014.

Limanowski, Jakub & Hecht, Heiko, "Where Do We Stand on Locating the Self?" Psychology. (Vol. 2 No. 4). July 2011. http://www.scirp.org/journal/PaperInformation.aspx?paperID=6284#.UyeMS61dXQE

Linsley, Alice C. "Is 'Good and Evil' a Biblical Merism?" Biblical Anthropology. http://biblicalanthropology.blogspot.com/2013/08/is-good-and-evil-biblical-merism.html

Lonergan, Bernard SJ. Insight: A Study of Human Understanding. (Volume 3, Collected Works of Benard Lonergan). Toronto. University of Toronto Press. 1992.

Long-Crowell, Erin. "Types of Heuristics: Availability, Representativeness, and Base-Rate." Study.com. https://study.com/academy/lesson/heuristics.html

Love, David K. Kepler and the Universe: How One Man Revolutionized Astronomy. Amherst, New York. Prometheus. 2015.

Loy, David. "Wei-Wu-Wei: Non-dual Action." Philosophy East & West. (Volume 35. Number 1.). 2019.

Lyne, William: Occult Ether Physics: Tesla's "Ideal Flying Machine" and the Conspiracy to Conceal It. Lamie, New Mexico. Creatopia Productions. 1997.

Madhavananda, Swami (trans.). The Brhadaranyaka Upanishad. Mayavati, India. Swami Yogeshvarananda Ashram. 1950.

Marcel, Gabriel. The Mystery of Being. Chicago. Henry Regnery Company. 1951.

Marris, Emma. "Where is Your Self?" Discover. January. 2015. http://discovermagazine.com/2014/dec/22-i-marks-the-spot

Marshall, Julia. "Image as Insight: Visual Images in Practice-Based Research." Studies in Art Education. 2007.

Matt, Daniel C. "Ayin: The Concept of Nothingness in Jewish Mysticism." In Robert Forman (ed.). The Problem of Pure Consciousness. Oxford University Press. 1990.

Matt, Daniel C. The Essential Kabbalah: The Heart of Jewish Mysticism. Castle Books. Edison, New Jersey. 1997.

Maxwell, James Clerk. "A dynamical theory of the electromagnetic Field." Philosophical Transactions of the Royal Society London. 1865.

McCraty, Rollin. "Heart-Brain Neurodynamics." Issues of the Heart: The Neuropsychotherapist (special issue) Dahlitz, M. & Hall, G. (Eds.). Dahlitz Media: Brisbane. 2015.

McDonald, J. H. (trans.) Tao Te Ching. Qigong Vacations.org. 2017.

McFadden, Johnjoe. Quantum Evolution: The New Science of Life. London. Harper Collins. 2000.

McGilchrist, Ian. The Master and His Emissary: The Divided Brain and the Making of the WesternWorld. New Haven.Yale University Press. 2009.

Mialet, Helene. Hawking Incorporated: Stephan Hawking and the Anthropology of Knowing. Chicago. Universtity of Chicago Press. 2012.

Michaelson, Peter. "When Eyes Are Blinders of the Soul." Why We Suffer.com. September 24, 2012. http://www.whywesuffer.com/when-eyes-are-blinders-of-the-soul/

Miller, Arthur I. Insights of genius: Imagery and creativity in science and art. Cambridge, Massachusetts. Massachusetts Institute of Technology. 1996.

Milutis, Jo. Ether: The Nothing That Connects Everything. Minneapolis. University of Minnesota Press. 2006.

Mlodinow, Leonard. Feynman's Rainbow: A Search for Beauty in Physics and in Life. New York. Warner Books. 2003.

Morelli, Sharon. "The Hara: Center of Being." Spirit of Change. http://www.spiritofchange.org/Spring-2011/The-Hara-Center-of-Being/

Morgan, Evan S. Tao, The Great Luminant: Essays from Huai Nan Tzu. United States. Library of Alexandria. 1933.

Moser, Barry (commentator). The New Revised Standard Version Bible with Apocrypha. Oxford. Oxford University Press. 1991.

Munro, J. Peter. "The Japanese Way of Tea is Other-Worldy." Design for Walking. https://designforwalking.com/the-japanese-way-of-tea-is-other-worldly/

Nagarjuna. J.L. Garfield (trans.) Madhyamakakarika. http://aaari.info/notes/03-06-06Tam2.pdf

Nagel, Ernest & Newman, James. Godel's Proof (Revised Edition). New York. New York University Press. 2001.

Nanamoli, Bhikku. The Buddha's Teaching in his Own Words. Kandi, Sri Lanka. Buddhist Publicaton Society. 1999.

Newberg Andrew & Lee, Bruce Y. "The Neuroscientific Study of Religious and Spiritual Phenomenon." Zygon. Volume 40, no.2. (2005)

https://static1.squarespace.com/static/52402ca4e4b0b7dd2fafe453/t/526049c6e4b034113969a71b/1382042054234/the-neuroscientific-study-of-religious-and-spiritual-phenomena-or-why-god-doesn%E2%80%99t-use-biostatistics.pdf

Newberg, Andrew et al. How God Changes Your Brain. New York. Ballantine Books. 2009.

Newberg, Andrew et al: Why God Won't Go Away: Brain Science and the Biology of Belief. NewYork, Ballantine Books. 2002.

Newberg, Andrew, et al. Words Can Change Your Brain, New York. Hudson Street Press. 2012.

Nietzsche, Friedrich. Ecce Homo; how one becomes what one is and The Antichrist. Thomas Wayne (trans.) New York. Algora Publishing. 2004.

Nietzsche, Friedrich. Human, All Too Human. Walter Kaufmann (trans.) Cambridge. Cambridge University Press. 1986.

Nietzsche, Friedrich. The Birth of Tragedy. Walter Kaufmann (trans.) New York, Random House.

Nietzsche, Friedrich. The Gay Science. Walter Kaufman (trans.). New York. Random House. 1974. 2010.

Nietzsche, Friedrich. The Will to Power. Walter Kaufmann (trans.). New York. Vintage Books. 1968.

Nietzsche, Friedrich: Thus Spoke Zarathustra. Walter Kaufmann (trans.). New York. Modern Library. 1995.

Oppenheimer, J. R. Quoted at "Vision '65: New Challenges for Human Communications" (October 1965) and published in v 65: New Challenges for human communications (Volume 4). International Center for the Typographic Arts. Southern Illinois University. 1965.

Oppenheimer, J. R. Science and the Common Understanding. New York. Simon & Schuster. 1953.

Penfield, Wilder. The Mystery of the Mind. New Jersey. Princeton University Press, 1975.

Parrott, Rodney J. The Samkhya Path of Observation.The University of Iowa. Dissertation. 1984.

Penrose, Roger. The Emperor's New Mind: Concerning Computers, Minds, and the Laws of Physics. Oxford. Oxford Universtiy Press. 1989.

Penrose, Roger. The Road To Reality. New York. Knopf. 2004.

Porter, William (Red Pine). Lao Tsu's Taoteching. Port Townsend, Washington. Copper Canyon Press. 2009.

Randall, Lisa. Knocking on Heaven's Door. New York. Harper Collins. 2011.

Randall, Lisa. "Theories of the Brane." Edge. April, 2018. https://www.edge.org/conversation/lisa_randall-theories-of-the-brane-lisa-randall

Randall, Lisa. Warped Passages: Unraveling the Mysteries of the Universe's Hidden Dimensions. New York. HarperCollins. 2006.

Reichenbach, Hans. The Rise of Scientific Philosophy. Berkeley, University of California Press, 1951.

Rickover, Robert. "Using Your Center of Gravity While Moving." Up with Gravity. September, 2012. https://www.upwithgravity.net/up-with-gravitysm-lesson-2-using-your-center-of-gravity-while-moving/

Ringu Tulku. Daring Steps Toward Fearlessness: The Three Vehicles of Tibetan Buddhism. Ithaca. Snow Lion. 2005.

Robinson, George. "What are the Sefirot?" My Jewish Learning. https://www.myjewishlearning.com/article/sefirot/

Rosenfeld, Arthur. "The Watcher Within You." Huffington Post. November 17, 2011. https://www.huffingtonpost.com/arthur-rosenfeld/the-watcher-within-you_b_181324.html

Rosenfield, Jordon. "Scientists are Trying to Solve the Problem of Awe." Science of Us. http://nymag.com/scienceofus/2016/05/scientists-are-trying-to-solve-the-mystery-of-awe.html

Rotman, Brian. Signifying Nothing: The Semiotics of Zero. Stanford: Stanford University Press, 1993.

Rucker, Rudy. Infinity and the Mind – the Science and Philosophy of the Infinite. Princeton. Princeton University Press. 2004.

Rumi, Jalaluddin.The Mathnawí of Jalálu'ddín Rúmí. (Vol. 4, part 7). Reynold Alleyne Nicholson (ed.). 1940.

Rumi, Jalaluddin. "Zikr". Creating Enlightened Awakenings. http://creawithin.com/blog/2012/06/23/zikr-by-rumi/

Sader, A L, Cha-no-yu: The Japanese Tea Ceremony. Rutland, Vermont. Charles E. Tuttle Company, Inc. 1933.

Sagan, Carl. Cosmos. New York. Random House. 1980.

Sagan, Carl. Varieties of Scientific Experience: A Personal View of the Search for God. New York. Penguin Books. 2006.

Salimpoor, Valorie M, "Anatomically Distinct Dopamine Release During Anticipation and Experience of Peak Emotion to Music." Nature Neuroscience. 2011. https://www.nature.com/articles/nn.2726

Sartre, Jean Paul. Being and Nothingness: An Essay on Phenomenological Ontology. (Hazel E. Barnes trans.) New York. Philosophical Library. 1956.

Sartre, Jean Paul. "Existentialism is a Humanism". Existentialism from Dostoyevsky to Sartre. (Walter Kaufman ed.) Ogden, Utah. Meridian Publishing Company, 1989.

Sartre, Jean Paul. Nausea. New York. New Directions. 2013.

Sartre, Jean Paul. No Exit and Three Other Plays. New York. Vintage Books. 1955.

Sartre, Jean Paul. Notebooks for an Ethics. (David Pellauer trans.) Chicago. University of Chicago Press. 1992.

Sartre, Jean Paul. The Words. New York. George Braziller. 1964.

Satler, Gail. Frank Lloyd Wright's Living Space. Decalb, Illinois. Northern Illinois University Press. 1999.

Seifert, M. "Calabi-Yau Compactification." Lecture of March 10, 2004. http://hamilton.uchicago.edu/~sethi/Teaching/P484-W2004/calabi-yau.pdf

Sekida, Katsuki (ed). Two Zen Classics: Mummonkan & Hekiganroku. New York. Weatherhill. 1977.

Sellars, Michael. "Heuristics for Good Hermeneutics." mechaelwsellars. May 24, 2012. https://wordoflifepublications.wordpress.com/2012/05/24/heuristics-for-good-hermeneutics/

Shankara. The Crest Jewel of Wisdom and other Writings of Shankaracharya. (Charles Johnston trans.) www.thefreedomreligionpress.org. 2012.

Shantideva. The Way of the Bodhisattva. (Padmakara Translation Group). Boston. Shambhala. 2006.

Shehan, Lawrence. "Physicists And Mystics: Similarities In World View." http://www.atpweb.org/jtparchive/trps-01-69-02-011.pdf

Shiota, Michelle N. et al. "The Nature of Awe: Elicitors, Appraisals, and Effects on Self-Concept", Cognition and Emotion. (21. 5). 2007.

Siderits, Mark & Katsura, Shoryu. Nagarjuna's Middle Way. Boston. Wisdom Publications. 2013.

Spade, Paul Vincent. "John Paul Sartre's Being and Nothingness." Lectures 1996. http://pvspade.com/Sartre/pdf/sartre1.pdf

St. John of the Cross. Dark Night of the Soul. Mineola, New York. Dover. 2003.

Susskind, Leonard. The Cosmic Landscape: String Theory and the Illusion of Intelligent Design. New York. Little, Brown & Co. 2006.

Susskind, Leonard."Using Maths to Explain the Universe; It can be hard to get beyond the limits of our neural architecture." The Economist. July, 2013.

Swiderek, Jolanta. "A Notion of Zero in the Philosophy of Aristotle." ΠΑΙΔΕΙΑ. https://www.bu.edu/wcp/Papers/Anci/AnciSwid.htm

Tanahashi, Kazuaki. The Heart Sutra: A Comprehensive Guide to the Classic of Mahayana Buddhism. Shambhala. Boston. 2014.

Tegmark, Max. Our Mathematical Universe: My Quest for the Ultimate Nature of Reality. New York. Knopf. 2014.

Tegmark, Max. "We're not insignificant after all." Less Wrong: A community blog devoted to refining the art of human rationality. http://lesswrong.com/lw/1li/max_tegmark_on_our_place_in_history_were_not/

Templeton, Graham. "Neil deGrasse Tyson says it's 'very likely' the universe is a simulation." ExtremeTech. April, 2016. https://www.extremetech.com/extreme/227126-neil-degrasse-tyson-says-its-very-likely-the-universe-is-a-simulation

The Holy Bible (Revised Standard Version). New York. Thomas Nelson & Sons. 1952.

Trungpa, Chogyam. Glimpses of Shunyata. Halifax. Vajradhatu Publications. 1993.

Tse-Tung, Mao. Quotations from Chairman Mao Tse-Tung. China Books & Periodicals. 1990.

Tversky, A & Kahneman, D., "Judgement under Uncertainty: Heuristics and Biases." sciencemag.org. https://science.sciencemag.org/content/185/4157/1124

Tyson, Neil deGrasse. Astrophysics for People in a Hurry. New York. Norton. 2017.

Tyson, Neil deGrasse. Death by Black Hole: And Other Cosmic Qanderings. New York. Norton. 2014.

Tyson, Neil deGrasse. "On the God of the Gaps." Moyers and Company. https://www.youtube.com/watch?v=ytaf30wuLbQ

Tyson, Neil deGrasse. "Universe Computer Simulation? - Neil DeGrasse Tyson Debate Emotional Speech." https://www.youtube.com/watch?v=aNoqWbsz8zs

Uskoković, Vuk. "The Metaphorical Model: The Bridge Between Science and Religion." Journal for Interdisciplinary Research on Science and Religion." (No. 6). January, 2010.

van der Eijk, Phillip. "Body and Spirit in Greek Medicine and Philosophy." Royal College of Psychiatrists. 2007. https://www.rcpsych.ac.uk/pdf/Philip%20van%20der%20Eijk%20Body%20and%20Spirit%20in%20Greek%20medicine%20and%20philosophyx.pdf

van Dongen, Jeroen. "Einstein and the Kaluza-Klein Particle." https://arxiv.org/pdf/gr-qc/0009087.pdf

van Elk, M., Karinen, A., Specker, E., Stamkou, E., & Baas, M.. "Standing in Awe: The Effects of Awe on Body Perception and the Relation with Absorption." Collabra (2.1. 4). http://doi.org/10.1525/collabra.36

Vey, Gary. "Where Are You Really?" Viewzone.com. 2011. http://www.viewzone.com/real.html

Volf, Carlo. "Light and the Aesthetics of Perception" Annual Conference of the Nordic Society for Aesthetics 2010. http://www.read.dk/ws/files/31582069/Annual_Conference_of_the_Nordic_Society_for_Aesthetics_2010_january.pdf

Watson, James. The Double Helix. New York. Atheneum.1968.

Watson, J.D., Crick, F.H.C. "A Structure for Deoxyribose Nucleic Acid". Nature (3). 1953.

Webb, Jeremy (ed). Nothing: From Absolute Zero to Cosmic Oblivion: Amazing Insights into Nothingness. London, Profile Books, Ltd. 2013.

Webb, Richard. "Nothingness: Zero, the Number They Tried to Ban." New Scientist. 22 November, 2011.

Weinberg, Steven. Dreams of a Final Theory: The Scientific Search for the Fundamental Laws of Nature. New York, Pantheon Books. 1994.

Whyte, Lancelot Law. The Unconscious Before Freud: A history of the evolution of human awareness. New York. Basic Books. 1960.

Wilczek, Frank. A Beautiful Question: Finding Nature's Deep Design. New York. Penguinn Books. 2015.

Wilczek, Frank. "Finding Nothing: A conversation with Noble Laureate Frank Wilczk." ASU Now.

https://asunow.asu.edu/20170208-finding-nothing-conversation-frank-wilczek

Wilczek, Frank. "Why 'Lightness of Being'?" http://www.lightnessofbeingbook.com/inside_Why.html

Wolfram, Stephan. "A Short Talk About Richard Feynman." Boston Public Library. https://www.stephenwolfram.com/publications/short-talk-about-richard-feynman/

Wright, Frank Lloyd. The Natural House. United Kingdom. Bramhall House. 1954.

Wyatt, C. S. & Schnelbach, Susan D. "The Existential Primer." http://www.tameri.com/csw/exist/

Yaden David B. et al. "The Overview Effect: Awe and Transcendent Experience in Space Flight." Psychology of Consciousness. (Volume 3, no. 3). 1989.

Yam, Philip. "Exploiting Zero Point Energy." Panacea BOCAF. http://www.panacea-bocaf.org/philipyam.htm

Yearley, Lee. In Experimental Essays on Chuang Tzu. Victor H. Mair (ed.). Asian Studies at Hawaii. 2010.

Young, Emma. "Gut Instincts. Secrets of Your Second Brain." New Scientist. December 2012.

Zarathustra. The Gathas, Our Guide. (Ali Jafarey trans.) California. Ushta, Inc. 1989.

Zhuangzi. Zhuangzi. Chinese Text Project. http://ctext.org/zhuangzi.